AF709353

VIKING REPRINT EDITIONS

THE Limitations OF SCIENCE

Q
162
.S85
1973

THE Limitations OF SCIENCE

by
J. W. N. Sullivan

BCL-75

AUGUSTUS M. KELLEY • PUBLISHERS
CLIFTON 1973

GOSHEN COLLEGE LIBRARY
GOSHEN, INDIANA

First Published 1933
(New York: The Viking Press, Inc.)
Copyright 1933 by The Viking Press, Inc.

RE-ISSUED 1973 BY
AUGUSTUS M. KELLEY · PUBLISHERS
Clifton New Jersey 07012
By Arrangement with THE VIKING PRESS

Library of Congress Cataloging in Publication Data

Sullivan, John William Navin, 1886-1937.
The limitations of science.

(Viking reprint editions)
Reprint of the ed. published by Viking Press, New York.
1. Science. 2. Science--History. I. Title.
Q162.S85 1973 501 77-122072
ISBN 0-678-03172-X

PRINTED IN THE UNITED STATES OF AMERICA
by SENTRY PRESS, NEW YORK, N. Y. 10013

Contents

THE Limitations OF SCIENCE

Introduction

SCIENCE, like everything else that man has created, exists, of course, to gratify certain human needs and desires. The fact that it has been steadily pursued for so many centuries, that it has attracted an ever-wider extent of attention, and that it is now the dominant intellectual interest of mankind, shows that it appeals to a very powerful and persistent group of appetites. It is not difficult to say what these appetites are, at least in their main divisions. Science is valued for its practical advantages, it is valued because it gratifies disinterested curiosity, and it is valued because it provides the contemplative imagination with objects of great æsthetic charm. This last consideration is of the least importance, so far as the layman is concerned, although it is probably the most important consideration of all to scientific men. It is quite obvious, on the other hand, that the bulk of mankind value science chiefly for the practical advantages it brings with it.

This conclusion is borne out by everything we know about the origin of science. Science seems to have come into existence merely for its bearings on practical life.

More than two thousand years before the beginning

of the Christian era both the Babylonians and the Egyptians were in possession of systematic methods of measuring space and time. They had a rudimentary geometry and a rudimentary astronomy. This rudimentary science arose to meet the practical needs of an agricultural population. Their geometry, a purely empirical thing,[1] resulted from the measurements made necessary by the problems of land surveying. The cultivation of crops, dependent on the seasons, made a calendar almost a necessity. The day, as a unit of time, was, of course, imposed by nature. The movement of the moon conveniently provided another unit, the month, which was reckoned from one new moon to the next. Twelve of these months were taken to constitute a year, and the necessary adjustments were made from time to time by putting in extra months.

This degree of scientific knowledge was the bare minimum necessary for the regulation of practical affairs. But another of the great motives for scientific research, disinterested curiosity, would seem to have played some part. The Babylonian priests continued to observe the heavens for long after their calendar had been established. They kept accurate records of the rising and setting of various heavenly bodies until, by the sixth century B.C. they were able to calculate in advance the relative positions of the sun and the moon, and so predict eclipses. The observations that were

[1] Recent discoveries show that the Babylonians had some grasp of theoretical principles.

made during the centuries that elapsed before this stage of perfection was reached could not have served any obvious practical purpose. They must have been undertaken out of curiosity, in order to discover what regularities existed amongst the motions of the heavenly bodies. But, once this degree of scientific knowledge had been reached, it was turned to practical account. Not, it is true, to the practical purposes of agriculture and the like, but to the no less practical purpose of foretelling the future in human affairs. Astronomy, in fact, was made to serve the purposes of astrology. Indeed, astrology was regarded as the real justification of astronomical researches.

There is nothing reprehensible in all this. It would show a grave lack of the historical sense to sneer at these early astronomers as being "superstitious." It must be remembered that the scientific outlook was not yet born. Science is not created by the scientific outlook; it is scientific knowledge that creates the outlook. In the time of these early Babylonian and Egyptian astronomers there was too little scientific knowledge in the world to justify them in creating a new outlook to accommodate it. They already had a comprehensive world-outlook, an outlook based on their experience and on their reasoning about it. They fitted the new facts into their general outlook, just as we do today. It was not until many centuries later that scientific facts became so abundant and recalcitrant that they obviously could not be fitted into the old outlook. Even as late as the seventeenth century so great a scientific man as Kepler used his astronomical knowledge to make astrological predictions—a little with his tongue in his cheek, perhaps.

We can see that a rudimentary knowledge of space and time measurements was imposed by the necessities of everyday practical life. Another science of obvious practical importance is the science of medicine. Medicine, as we should expect, is one of the oldest of the sciences. But here the general Babylonian outlook on life put them at a marked disadvantage. Their experience of life had convinced them that the universe is governed by powers that are, on the whole, maleficent. It seemed to them that pain and disease could well be referred to the direct action of the gods. They therefore had recourse to sorcery and exorcism as the only way of dealing with the problem. Rational medicine made no progress whatever. Life in Egypt was more secure, was less liable to sudden storms and floods, and the universe appeared to the Egyptians as a less malignant affair. Their mythology shows the divine powers as being, for the most part, friendly to man. They practised incantation in their treatment of diseases, but they also looked for other causes than the direct action of the gods. And their practice of embalming their dead gave them some knowledge of anatomy. Egyptian medicine reached a considerable degree of development.

In none of these discoveries does there seem to have

been more than a trace of what is called the scientific spirit. The only scientific problems that interested these ancient peoples were those that had a direct bearing on practical affairs. They seem to have shown little, if any, disinterested curiosity in the workings of nature. And they based no speculations on their scientific discoveries. These discoveries were incorporated into their religions and philosophical schemes and were interpreted in accordance with their religious and philosophical principles.

It is not until we reach the Greeks that we find science emerging as an autonomous activity. It is not until then, in fact, that we find anything that we can call the scientific spirit. Thales of Miletus (*c.* 580 B.C.), we are told, set out to answer the question, "Of what, and in what way is the world made?" Here we recognize the spirit, necessary to science although not peculiar to it, of disinterested curiosity. The Greeks appear to have been the first people with whom this feeling became a passion. They wanted to know—for the sake of knowing. All their predecessors, it seems, like so many of their successors, belonged to the type which asks, "What is the use of it?" It really seems as if the human consciousness, with the rise of the ancient Greeks took a genuine leap forward. An unexampled freedom of the mind was born. This was a necessary condition for science to come into the world.

In another respect, also, the Greeks were unique. They seem to have been the first people with a thorough grasp of the nature of mathematical reasoning. The land-surveying formulae of the Egyptians gave rise, in the hands of the Greeks, to a deductive geometry. This was an immensely important step forward. Mathematical reasoning, the most powerful of man's intellectual instruments, was created. Overwhelmed by the almost magical power of this new instrument, the Greeks thought that in mathematics they had discovered the key to all things. To the Pythagoreans, in particular, number was the principle of all things. Everything, whether physical properties or moral qualities, was a manifestation of number.

This outlook has played a very large part in the development of science. Leonardo da Vinci's remark that a science is perfect in so far as it is mathematical has been very generally accepted by scientific men. If a complete mathematical description of the world could be given, it is felt that science would be complete. But is there any *a priori* reason to suppose that the universe *must* be the kind of thing that can be described mathematically? To Newton, at any rate, the attempt to describe nature mathematically was an adventure that might or might not be successful. And some modern men of science have been so astonished by the success of the adventure that they have been led to conclude that God must be a mathematician. On the other hand, there seems some reason to believe that any universe containing several objects can be brought within some sort of mathematical web, so that the mathematical character of the universe is a fact of no particular significance.

But whatever basis of truth there may be in the Pythagorean outlook it is certain that it greatly exaggerated the significance of mathematics. Nevertheless, a modified form of this outlook, after many fruitless centuries had elapsed, contributed very powerfully towards the origin and development of the modern scientific movement.

In the meantime the spirit of disinterested curiosity and man's delight in this new and wonderful mathematical faculty, withered and died under the cold blight of the Roman Empire. The Romans were an essentially practical people, and they adopted the What is the use of it? attitude towards all abstract speculation. Such science as they had was borrowed from the Greeks, and they seem to have valued it solely for its practical applications in medicine, agriculture, architecture, and engineering. As a natural consequence of their obsession with practical affairs the Romans created nothing in science.

The ensuing centuries in Europe, up to the time of the Renaissance, also produced nothing in science. But this was not because the medievalists were exclusively absorbed by practical affairs. On the contrary, some of the greatest abstract thinkers the world has ever produced appeared at this time. But they had an outlook on life that made science unnecessary. Science could tell them nothing that they wanted to know, and they had no curiosity about the sort of things science could tell them. The medievalist lived in an orderly universe. He knew the principles on which it was constructed, and he knew the meaning and purpose of everything in it. He knew the scheme of creation; he knew the end that every created thing was made to serve. He derived this information from two sources, reason and revelation. The highest discoveries of the human reason were embodied in the works of Aristotle; the Scriptures contained divine revelations on matters not accessible to reason. By synthesizing these two kinds of information everything worth knowing could be learned. This synthesis was accomplished, magnificently, by St. Thomas Aquinas.

The medievalist lived in a purposeful universe of which he himself was the centre. The reason why phenomena existed was to be found in their bearing on the eternal destiny of man. Nothing had any meaning except in so far as it fitted into this great logical scheme. In this atmosphere it is obvious that science would appear to be a trivial activity. It could be of no real importance, for the reason that it was concerned with merely secondary questions. *How* things happened was of no importance compared with the question of *why* they happened. Even Roger Bacon, the one man of his time who insisted on the experimental investigation of nature, agreed that the importance of this investigation was that it would assist in elucidating theology. It was only when faith in the all-pervading purposefulness of natural phenomena had faded that the scientific method of inquiry became important.

But although the scholastic outlook discouraged scientific inquiry, it furnished an essential element of the scientific outlook itself. This was the belief in nature as a rational whole. In the medievalist's universe, unlike that of the Babylonians and other early peoples, nothing was capricious or arbitrary. This belief, that "every detailed occurrence can be correlated with its antecedents in a perfectly definite manner, exemplifying general principles" is, as Whitehead says, the necessary basis for the whole scientific adventure. "Without this belief the incredible labours of scientists would be without hope." Yet this belief in universal order does not impose itself as an outcome of direct experience, as the very different conceptions prevalent in earlier times is sufficient to show. It may even be that this belief will ultimately prove to be unjustified. It may be, as Eddington has hinted, that the universe will turn out to be finally irrational. This would mean, presumably, that science would come to an end. This does not mean, of course, that the scientific knowledge so far obtained would be abandoned. As a set of working rules science would still be valid, for phenomena would presumably continue to occur in the same fashion as at present. But science would have reached a limit beyond which it could not go.

The development of science up to now, then, has assumed that nature is a rational whole, and this belief we owe, as a matter of history, to the great scholastic philosophers. Although, therefore, they achieved nothing, or practically nothing, in actual scientific discovery, they had a great deal to do with the formation of the modern scientific outlook.

That outlook comes to its first clear expression in Galileo. During the great intellectual ferment of the Renaissance a scientific genius of the first order appeared in the person of Leonardo da Vinci, but unfortunately he never published his scientific researches. What influence he may have had on the succeeding century could have been only indirect. And even Copernicus, immensely important though his work was, did not so completely manifest the scientific spirit as did Galileo. Copernicus was led to his assertion that the earth and the other planets went round the sun chiefly by considerations of mathematical harmony. The Copernican system was, regarded mathematically, a very much neater affair than the Ptolemaic system that it replaced. It was, however, open to objections that were at that time unanswerable. Also, it was in conflict with the general outlook of the time, which still regarded man as the centre of the universe. Nevertheless, its æsthetic charm, considered as a mathematical theory, was sufficient to secure it the enthusiastic acceptance of such rare spirits as Galileo and Kepler. They felt that so beautiful a thing must be true although, as Galileo admitted, it seemed to contradict the direct testimony of our senses.

Even Galileo himself was not the *perfect* scientific man. Perfection was reached only in the person of Isaac Newton. Galileo fell a little short of the possible by not fully realizing the necessity of confirming mathematical deductions by experiment. Fortunately, the objections of his opponents forced him to make test experiments.

This tendency to rest content with the mathematical deduction has always been characteristic of a certain type of scientific man, and was particularly noticeable at the beginning of the scientific movement. In the case of Kepler this tendency was supported by a whole philosophy. Kepler believed that the very reason for phenomena being as they are was that they fulfilled certain mathematical relations. By discovering these mathematical relations we seize upon the purpose that guided the Creator.

But although Kepler's philosophy led him into innumerable fantastic speculations, he was always stubbornly faithful to the facts. His anguish at finding that some wild and beautiful idea was not confirmed by observation was, as we know, sometimes very considerable, but he never hesitated to abandon it. He was spurred on, indeed, to look for an even more subtle and recondite harmony. And he succeeded in finding it. His three laws of planetary motion are not only of the first importance scientifically, they are also beautiful. And this quality of his imagination led him also to exceptionally beautiful ideas in the realm of pure mathematics. Kepler, more than any other man, conveys to us the breathless excitement that must have attended the opening of the great scientific movement. The poetry of science and its sense of unlimited adventure are conveyed by Kepler in the most magnificent prose that any scientific man has ever written.

When we come to Newton the sun is fully up. The scientific outlook has, in him, reached full consciousness. It would be fair to say that science, in the hands of Newton, has become a completely autonomous activity for, although Newton had a philosophy and a religion, they did not play any part in his science. The basis of science, according to Newton, was observation and experiment. From this basis mathematical deductions could be made. These deductions were then to be checked by further experiment. Thus science formed an independent and self-enclosed system, borrowing nothing, as it had done formerly, from metaphysics or theology. This outlook was not understood by Newton's contemporaries. It was, as it were, too austere for them. But it has become the dominating outlook of the scientific world.

CHAPTER I

The Expanding Universe

In discussing the extent to which science has met our curiosity about the universe, we must remember that different minds are curious about very different things. It is reported that once when Arago, the great French astronomer, was expounding the nature of comets at a dinner party, Victor Hugo, who had been listening to the exposition with a baffled expression, said finally: "But, Monsieur Arago, what is the *soul* of a comet?" Arago's own baffled expression at this question revealed the essential dissimilarity of the two minds. There are, indeed, many minds for whom science is of no interest whatever. Their curiosity is exclusively concerned with questions that science is in no position to answer. We have seen that the scholastics were, for the most part, in this position; their outlook on the world made science seem uninteresting. At the present day it is possible to meet people who regard science as trivial, as throwing no light on the problems which most concern mankind. Even such earnest and inquiring thinkers as Tolstoy and Dostoievsky could not see that science was more than a comparatively insignificant activity. And we may suspect that amongst those who profess to value science there are some who value it only for its practical applications. Nevertheless, the history of science makes it clear that it has, on the whole, been prosecuted, not for its practical advantages, but for its power of meeting man's curiosity about the universe. The earliest, and also the most important of the sciences, from this point of view, is astronomy.

The early speculators on the structure of the universe reached their conclusions on the basis of what they thought probable—as we do today. But a man's estimate of probabilities varies with his experience. The idea, for instance, that the earth is a mere speck in the immensity of space is not an idea that would naturally occur to anyone. Judged by the direct testimony of the senses, it is by far the largest thing in man's experience. The old observers, who thought that the sun and moon were moderate-sized bodies, situated a moderate number of miles from the earth, were perfectly justified. Modern estimates of these distances would, rightly, have appeared to them as incredibly fantastic. And it was perfectly sensible to suppose that these bodies did what they appear to do, namely, circle round the earth. Indeed, these ideas were not reached except by a great effort of abstraction. For they assumed that the earth remained unsupported in the middle of space. More primitive thinkers had assumed, again quite naturally, that the earth was supported, and they imagined a variety of exceedingly fantastic supports for it such as monstrous tortoises, elephants, and so on. The idea of an isolated and immobile earth, with the heavenly bodies circulating round it, was a great advance. As we have seen, it commended itself to a great number of first-class minds.

It so happened, however, that this theory turned out to be not so simple as it seemed. It fits in well enough with rough-and-ready observations. A superficial glance at the sky confirms it. But amongst the circulating bodies were observed a few—the planets, or wanderers—whose motion, when carefully studied night after night, was found to be not regular. To fit these into the scheme required considerable ingenuity. As instruments improved, and observation became more precise, the ingenuity necessary to account for these irregularities became ever greater. For centuries astronomers, wedded to the principle of circular motion, spread ever great complexity throughout the heavens. Now complexity is nearly always unæsthetic, and by the time Copernicus appeared the heavens, when regarded narrowly enough, were an unæsthetic jumble of arbitrary and varying motions. Copernicus had the æsthetic tastes proper to a mathematician, and this state of affairs seemed to him intolerable. It seemed to him that nature could not be as complicated and ugly as all that, and he began to wonder whether some simpler explanation of these puzzling notions could not be given. He found that some of the ancients had held that the earth moved, and he meditated upon this possibility. He found, when he came to work out this idea mathematically, that a far more harmonious system of the universe emerged.

This was sufficient. The fact that the idea was opposed to the prevailing philosophy, and also opposed to the testimony of the senses, was not sufficient to overcome the belief that so beautiful an idea must be true. Copernicus published his theory, confident that its æsthetic charm was its sufficient justification. As he says: "We find, therefore, under this orderly arrangement, a wonderful symmetry in the universe, and a definite relation of harmony in the motion and magnitude of the orbs, of a kind it is not possible to obtain in any other way." His confidence that mathematicians, at any rate, would find his system irresistible, was justified. Kepler's outburst is characteristic. "I have attested it as true in my deepest soul," he says, "and I contemplate its beauty with incredible and ravishing delight." Other people, however, were not convinced until after Galileo's invention of the telescope and his application of it to the survey of the planets. He showed that the inner planets, Venus and Mercury, exhibited phases just as was required by the Copernican theory, and his telescopic view of Jupiter, with its four moons going round it, showed a system analogous to the Copernican idea of a sun with circulating planets.

Men were forced to expand their notions of probability to fit this new evidence. It became definitely accepted that the earth is not at the centre of the universe. But other, and equally unlikely results, followed from this new fact. For, if the earth moves in some tremendous orbit round the sun, the stars should seem to shift against the sky, just as trees, seen from a passing train, seem to shift their position against the horizon. This effect was looked for, but could not be found. Men were reluctant to draw the startling conclusion to which this negative result pointed. For this negative result could only be explained by supposing that the stars are at simply incredible distances, very much farther than any that man had ever imagined. And, as has been plausibly suggested, men's notions of probability, at that time, were hampered by farming considerations. Such enormous stellar distances pointed to an immense waste of space, and it is probable that the men of that time, even if unconsciously, judged God's notions of economy as they would judge a farmer's. But by the middle of the nineteenth century the distance of a star was actually determined. The finest instruments of that time were just sufficient to determine the shift of some of the nearer stars. With this discovery the ground was cleared for those dazzling researches which have led to our present notions of the scale of the universe.

§ 2

The distances, ages, temperatures, etc. dealt with by modern astronomy are quite unimaginable and are, in a sense, uninteresting. When man believed that his significance in the universe was in some way dependent on his physical size, or on the physical size of his earth, he would doubtless have been somewhat disconcerted to learn that the universe was ten times bigger than he had thought it was. Having accommodated himself to this new notion he might have received another shock, although a fainter one, to find that the previous result had erred on the side of moderation, and that the size of the universe must be again multiplied by ten. But there is a limit to this process. There is not much difference, psychologically, between realizing that one is a drop in a bucket and realizing that one is a drop in an ocean. Astronomy, in its aspect of revealing man's physical insignificance in the universe, reached this stage long ago. The latest estimates merely turn the bucket into an ocean.

The farthest bodies that can be observed with our present instruments are distant from us by well over one hundred million light years.[1] It is to be expected that further instrumental power will reveal still more distant bodies. It might be thought that this process could be continued indefinitely, that, just as modern astronomy reveals a larger universe than that known to the Victorians, so the astronomy of the future will reveal a larger universe than that known to us. But we have reason to believe that there is a natural limit to this process.

---

[1] That is to say, light, which travels 186,000 miles in a second, would take over one hundred million years to reach these bodies.

The greatest change that has occurred in our way of thinking about the universe since the days of Copernicus is the modern conception of the universe as finite. This does not mean merely that there is a finite amount of matter in space, but that space itself is finite. The instant impulse, that most laymen have, to reject this statement as obviously absurd, is really an indication that they are unconsciously dominated by certain assumptions. These assumptions can be isolated and analysed, and shown to be logically arbitrary. The analysis is now about a hundred years old. It began with the invention of the first non-Euclidean geometry.

About one hundred years ago Lobachevsky, a Russian, and Bolyai, a Hungarian, found, independently of one another, that Euclid's geometry is not a logical necessity. This is not to say that Euclid's geometry is incorrect. If we start by accepting Euclid's axioms, then the whole superstructure Euclid built up on them must be accepted also. But the fact is that it is not necessary to accept Euclid's axioms. They are not "necessities of thought." Quite different axioms can be postulated and a perfectly self-consistent system of geometry built up on them. It then becomes a matter of experiment to determine which system of geometry best fits our experience.

The invention of the non-Euclidean geometries is one of the most remarkable feats in the intellectual history of man. For two thousand years Euclid's axioms had reigned unchallenged. That they were "necessary truths," true for angels as well as men, true even for God Himself, was admitted by all the philosophers. Merely to wonder whether these truths could be transcended, merely to wonder whether there was a world of ideas outside them, was an effort of extraordinary imaginative daring. And to translate this scepticism into the creation of a new, coherent, and complete system of geometry was a wonderful and exhilarating achievement of the free human mind. In this achievement the human intellect had reached a new level of abstraction; there was a veritable growth of consciousness. Of course, so stupendous an achievement was not at first understood. Even the mathematicians of the time thought that Lobachevsky and Bolyai were mad. And the great Gauss, who had in private reached similar conclusions to theirs, confessed that he had been afraid to publish his discovery. But gradually the new ideas were assimilated. Gradually the human imagination learned to expand to these new concepts. A new mathematical era had dawned.

For our present purpose the most important thing about this development was that the science of deductive geometry acquired a new status. It became clear that we can start with any set of axioms we please, provided they are consistent with one another, and on the basis of these construct a geometry which is, logically, as impeccable as Euclid's. Some of these geometries are very like Euclid's, and some of them are very queer indeed. Now the question arises, which of these geometries are we to apply to actual physical space, the space in which the sun and planets and all the millions of stars exist? As long as it was assumed that the only possible geometry was Euclid's geometry, we had, of course, no choice. But now that we know that other geometries are possible, we can ask the question. The question is to be decided by measurement. Our measuring instruments may be rods and clocks, or rays of light and vibrating atoms. What geometry best describes their observed behaviour? It is found, when the measurements are made, that a particular form of geometry, invented by the German geometer Riemann, best meets the facts of the case. Einstein was the first to realize this, and he has applied this geometry with immense success. Scientific men are now convinced that we live in a space which is not governed by the laws of Euclid's geometry, but by the laws of Riemann's geometry.

This discovery has, from the common-sense point of view, some very singular consequences. For instance, the total volume of space is finite. This does not mean that space has any boundaries. It does not stop short anywhere. Nevertheless, it is not unlimited.

This is a very singular idea, but fortunately we can illustrate it by an example drawn from ordinary experience. Everybody knows that a volume has three dimensions, length, breadth, and thickness, whereas a surface has only two, length and breadth. Now it is quite possible, for the purposes of argument, to imagine creatures who live only in two dimensions. They never look up or down; they know nothing of up or down. They are conscious only of length and breadth. Now, if these creatures lived on a flat surface, like the surface of a table, they would find that their space (the surface of the table) could not be unbounded unless it was also infinite. They could only go on in a straight line for ever provided the table went on for ever. Unless the table was actually infinite in length and breadth, they would sooner or later reach its edges and find that it had boundaries. They could not continue to go on without leaving the surface of the table altogether. Now consider the surface of a sphere. Here the creatures could crawl about indefinitely, sticking to the surface of the sphere, crawling in any direction, and however long they kept on they would never come to a boundary. Yet the total area of their space (the surface of the sphere) is finite. Thus we see that, when we confine ourselves to two dimensions, at any rate, it is possible to have a space which is finite but unbounded.

Now it is significant that the geometry of a flat surface is Euclidean, whereas the geometry of a spherical surface is not. According to Euclid the three angles of a triangle add up to two right angles. This is true if we draw triangles on the surface of a table or on a blackboard. It is not true if we draw triangles on a sphere or on an egg. And many others of Euclid's theorems are wrong if we draw our diagrams on a sphere. If our two-dimensional creatures on the sphere were mathematicians, they would never dream of inventing Euclid's geometry. The geometry they would invent, the one that best fitted the facts of their world, would be one of the non-Euclidean geometries called spherical geometry. It would only be when a Lobachevsky appeared amongst them that they would think of anything so fantastic as Euclid's geometry.

Now it so happens that the Riemannian geometry that Einstein adopted is a glorified sort of spherical geometry. It shows that a three-dimensional space can have the same odd properties as a two-dimensional spherical space, that is, it can be finite but unbounded. And mathematicians are now convinced that this is the sort of space we live in. In this space we could go on in a straight line for ever without ever meeting any boundary to our further progress. But we should find, when we had been going on for sufficiently long, that we had come back to our starting point. Space is finite but unbounded.

§ 3

What is the volume of this space? Recently a very curious answer has been given to this question. Space has no definite volume, for it is continually expanding!

"We can picture the stars and galaxies," says Professor Eddington, "as embedded in the surface of a rubber balloon which is being steadily inflated; so that, apart from their individual motions and the effects of their ordinary gravitational attraction on one another, celestial objects are becoming farther and farther apart simply by the inflation. It is probable that the spiral nebulæ are so distant that they are very little affected by mutual gravitation and exhibit the inflation effect in its pure form. It has been known for some years that they are scattering apart rather rapidly, and we accept their measured rate of recession as a determination of the rate of expansion of the world."

It is supposed that, some thousands of millions of years ago, the matter of the universe was more or less in a state of equilibrium. At this period the radius of space is estimated to have been something over one thousand million light years. But the state of equilibrium was unstable. The universe had an inherent tendency to expand which was only approximately counteracted by the gravitational attraction of the matter in the universe. Once the expansion began, however slowly, it was bound to go on, for as the matter of the universe became more scattered its gravitational attraction became less effective. The radius of space today, Professor Eddington estimates, is not less than ten times its original radius. And the actual rate of expansion is itself continually increasing. It is probable that the radius of space is already increasing faster than the velocity of light, and this rate of expansion will grow greater. Already it would be impossible for a ray of light to go all round the world, for the circumference of the world is growing faster than light could overtake it. In the original state of equilibrium light travelled round and round the world until it was absorbed. Then the universe began to expand, and there came a period when the rate of expansion made it impossible for light to complete its journey. At this period "the bubble burst." It has been suggested that some of the nebulæ we see in the heavens are seen from the back, by light which has travelled all round the universe. If this is so, then we are seeing these nebulæ by light which started on its journey before the rate of expansion of the universe reached the critical value—before the bubble burst, in fact.

This theory was introduced into science by the mathematicians, but there is good observational evidence for it. The vast majority of the nebulæ whose motions have been measured are moving away from us with very great speeds. And these speeds are greater the farther the nebulæ are from us, which is just what we should expect on the mathematical theory.

The spiral nebulæ, whose rate of scattering enables us to determine the rate of expansion of the universe, are the largest and the most isolated bodies in space. Indeed, they are often referred to as "island universes." It is estimated that, on the average, each of these "island universes" contains enough matter to make many thousands of millions of stars, each the size of the sun. And the sun, it must be remembered, is about one million times bigger than the earth. About two million of these nebulæ are visible in the great 100-inch telescope at Mount Wilson observatory. There is every indication that an increase in telescopic power would reveal more of them—probably in proportion to the increase of telescopic power, as they show no signs of thinning out. They are truly "island universes," for their average distance apart is of the order of two million light years.

Our sun is a member of one of these systems.

In so gigantic a universe it is not surprising to find an immense variety. In the different bodies constituting the universe the range of size, of density, of temperature, is exceedingly great. The light and heat emitted by the stars is, in some cases, many thousands of times less than that of the sun and, in some cases, many thousands of times greater. There are stars of an inconceivable tenuity, and there are stars of a density far surpassing all earthly standards. There are stars no larger than the earth, and there are stars millions of times bigger than the sun. And this universe, in its present condition, is certainly thousands of millions of years old, and it may be that its age must be reckoned even in millions of millions of years.

§ 4

So far as size goes, the universe, it would appear, can have no surprises left for us. The figures we have given cannot impress us more, even if they come to be multiplied by a hundred or a million, for they are already unimaginable. The physical insignificance of man and his planet is amply demonstrated. It is henceforth impossible to make man's importance dependent on his physical dimensions or position, as the medievalists did.

The significance we attach to this fact depends on our philosophy. There are those who find the mere size and age of the material universe irrelevant to the question of man's importance. There are others who think that life is an accidental, meaningless, and extremely fleeting episode in the history of matter. Even if we suppose that there are other intelligences than ours, existing on other planets, it is practically certain that life and consciousness must be very rare phenomena. According to the accepted theory, which supposes that planets are produced by the chance close approach of two stars, there may not be more than one star in every hundred thousand which possesses a planetary system. And it could only be rarely, even then, that any of these planets would be fit to support life—at least, as we know it. In our own system, for instance, there is no good evidence that any planet but our own could support life. Mars is a possible exception but, for the majority of the planets of the solar system, we can say definitely that life on them is impossible. Even so, however, the number of stars in the universe is so enormous that, assuming that life is the chance outcome of the combination of certain physical conditions, we should have to assume that there are some millions of inhabited planets.

There is reason to suppose, also, that of these inhabited planets ours is one of the youngest. If the evolution of intelligence has proceeded everywhere on similar lines to our own, we can say that the greater part of the living beings in the universe are immensely more developed than we are. It would be impossible, even on this evidence, to say definitely that life is not the purpose of the universe. For we know, from ordinary observation, how exceedingly wasteful and haphazard are nature's means of securing her ends. For every seed that comes to life thousands die. There is nothing opposed to what we know of nature's economy in supposing that a million stars burn themselves out through millions of years in order that a few brief and evanescent lives may appear. We cannot apply to the universe notions of probability based on human behaviour.

Nevertheless, it would never occur to anybody, surveying what science has to tell about the past and probable future of the universe, to suppose that the production of life and consciousness was the purpose of the process. But the scientific account of the process is not yet completely coherent. One of the least disputable laws of physical science states that the universe is steadily running down. The energy of the universe, although it remains constant in amount, is steadily assuming a less available form. The stars are continually radiating energy, which may become dissipated throughout the whole of space, but which never returns in its original concentrated form. All the matter in the universe must, finally, either dissipate itself as energy, or else assume one dead level of temperature. All energy interchanges will then be impossible. A final state of equilibrium will have been reached and, in this unchanging condition, the condition of "maximum disorganization," the universe will persist indefinitely. This process, it must be remembered, is taking place now. Except for the few rare corners where more complex organic beings are being built up, we live in a wasting universe. And the small local building-up processes must finally be overwhelmed by the general tendency towards disorganization.

But the fact that the energy of the universe will be more disorganized tomorrow than it is today implies, of course, the fact that the energy of the universe is more highly organized today than it will be tomorrow, and that it was more highly organized yesterday than it is today. Following the process backwards we find a more and more highly organized universe. This backward tracing in time cannot be continued indefinitely. Organization cannot, as it were, mount up and up without limit. There is a definite maximum, and this definite maximum must have been in existence a finite time ago. And it is impossible that this state of perfect organization could have been evolved from some less perfect state. Nor is it possible that the universe could have persisted for eternity in that state of perfect organization and then suddenly, a finite time ago, have begun to pursue its present path. Thus the accepted laws of nature lead us to a definite beginning of the universe in time. We are to suppose, on this reasoning, that, at some particular moment in the past, a perfectly organized universe sprang suddenly into being, and has been steadily becoming more and more degraded ever since.

There is nothing logically impossible in this conclusion, but it nevertheless seems to be utterly incredible. It is highly probable that we are here faced with a real limitation in the scientific scheme. The analysis on which these results are based has not, we may suppose, been sufficiently discriminating. It may be, for instance, that the notion of *time,* as it occurs in science, is too meagre an abstraction. It may be that it is due to the inadequacy of this abstraction that the doctrine of the decaying universe leads us back to an actual beginning of time. Or it may be that the scientific scheme has wholly omitted some essential element. This is Professor Whitehead's view when he says:

> The moral to be drawn from the general survey of the physical universe with its operations viewed in terms of purely physical laws, and neglected so far as they are inexpressible in such terms, is that we have omitted some general counter-agency. This counter-agency in its operation throughout the physical universe is too vast and diffusive for our direct observation. We may acquire such power as the result of some advance. But at present, as we survey the physical cosmos, there is no direct intuition of the counter-agency to which it owes its possibility of existence as a wasting finite organism.

There is some evidence, not very convincing, of a

building-up process in the universe. It has been known for some years that a highly penetrating radiation is reaching the earth from outer space. Observation shows that it does not come from the sun, and, as the sun is a typical star, there is therefore no reason to suppose that it comes from the stars. It probably comes from regions quite outside the particular stellar universe, the particular spiral nebula, to which we belong. Respecting the origin of this radiation there are two theories, which we shall discuss in more detail when we come to consider the nature of matter. Here we shall say briefly that one theory holds that the radiation is produced by the building up of matter, whereas the other theory holds that it is produced by the annihilation of matter. Each of these processes is theoretically possible, but the second seems to be more likely.

Surveying the universe presented by modern astronomy we see that the vast majority of the bodies in it seem to exist without any reference to the needs of life and consciousness. The vast extent of the universe, both in space and in time, is, from the human point of view, completely aimless. Those immense lumps of matter, in their millions of millions, incessantly pouring out an inconceivably furious energy for millions of millions of years, seem to be completely pointless. For a fleeting moment man has been permitted to stare at this gigantic and meaningless display. Long before the process comes to an end, man will have vanished from the scene, and the rest of the performance will take place in the unthinkable night of the absence of all consciousness. This revelation is startling. It is still more startling, almost incredible, when we reflect that this amazing panorama sprang suddenly into existence a finite time ago. It emerged full-armed, as it were, out of nothing, apparently for the sole purpose of blazing its way to an eternal death. This is the scientific account. It seems to be true as far as it goes, but we cannot believe that it is the whole truth. We prefer to believe that the present scientific method has its limitations.

CHAPTER II

The Mystery of Matter

A VERY considerable knowledge of the properties of matter was obtained in prehistoric times. The discovery and use of fire introduced man to a whole range of new phenomena. We know, for example, that cooking, the fermenting of grape-juice, the smelting of metals, and the making of stoneware were prehistoric achievements. In ancient Egypt the arts and crafts had reached a considerable level of development. The Egyptians made pigments and cosmetics from metallic compounds; they were skilful in dyeing; they could temper iron; and they could make glass and enamel. To accomplish all this they must have had a very considerable practical knowledge of both physics and chemistry. But here, again, it appears that the Greeks were the first people to speculate about the theoretical implications of their knowledge. Thales observed that there is a cycle in nature, a cycle of air, earth, and water through the bodies of plants and animals to air, earth, and water again. He also observed that the food of plants and animals is moist, and was led to conclude that all things are composed of water in various stages of condensation or rarefaction.

The most obvious thing about the world of matter is its immense variety. If we look abroad on a landscape, containing water, clouds, trees, soil, rocks, grass, flowers, our own bodies, and our clothing, we may well wonder how to classify so great a range of material. The Greeks seem to have been impressed chiefly by the densities and forms of the various kinds of matter. Thus they classified matter into four elements which, given in order of decreasing density, are earth, water, air, fire, for they regarded fire as a rarer and more refined form of air.

Since the essential difference between these elements was regarded as one of density, it was natural to take one of them as primordial and to derive the others from it either by a process of condensation or by one of rarefaction. Thales, as we have said, chose water as the fundamental element. Anaximenes chose air. Heraclitus chose fire, but a very refined sort of fire, an ethereal fire, a kind of soul-stuff. It was also possible, of course, to maintain that all four elements existed in their own right, as it were. The Sicilian philosopher Empedocles worked out the theory that all existing varieties of matter are formed by combinations of the four elements in different proportions, these proportions combining under the influence of two contrasting divine powers, one of attraction and one of repulsion. In human beings these contrasting powers manifest as love and hate.

A specious demonstration of the theory that complex substances consist of the four elements is given by burning green wood. Burning, according to the Greeks, resolves a substance into its component elements. Now, when a piece of green wood is burned, we see fire, we see smoke that passes into air, we see water bubbling from the ends, and we are finally left with an ash that is obviously earthy in character. This demonstration owes its plausibility, of course, to the fact that it rests on totally incorrect ideas of the nature of combustion. But correct ideas on this subject were not obtained for very many centuries. The scientific movement was in a state of vigorous maturity before this problem was solved.

A more important scientific product of Greek thought was the Democritean theory of atoms. What happens when a piece of matter is continually divided and subdivided? Does it preserve all its qualities intact? Thus, is a very tiny piece of green leaf itself green? Is a very tiny piece of sugar sweet? A certain school of Greek thought maintained that the qualities or properties of matter were ultimate things about it. However far the process of division might be carried, earth would remain earth and water water. Thus the attempt to reduce matter to simpler elements was, in effect, declared impossible on this view. Democritus was of the contrary opinion. "According to convention there is a sweet and a bitter," he says, "a hot and a cold, and according to convention there is colour. In truth there are atoms and a void." All atoms, he taught, are of the same substance, although they are of various sizes and shapes. The differences of the different kinds of matter are reduced to the differences of size, shape, and motion of the atoms that compose them.

This theory, in principle, is not unlike the modern theory, although there are very important points of dissimilarity. The theory was not fruitful scientifically since it was not given a quantitative form, and could not be used to explain phenomena in detail nor for prediction. But it was entirely in harmony with modern scientific thought in making the "real" world of matter something entirely different from the vivid coloured world perceived by the senses. The conception of the real world as a vast machine, colourless, odourless, soundless, had been introduced into human thought.

We see again the difficulty of hitting on a satisfactory classification of material substances when we come to the alchemists, the precursors of modern chemistry. The early alchemists attached immense importance to colour. They believed that any metal that could be made to look like gold, for instance, was thereby transformed into gold. For them all was gold that glittered. This belief was part of a general outlook. On the basis of the Platonic philosophy it was generally accepted that the "reality" of anything was to be found in the ideals that it embodied. The mere stuff of matter was regarded as unimportant compared with its qualities, just as a man is to be valued by the nature of the soul that informs him and not by the mere stuff of his body. Just as the heavenly bodies, in astrology, are embodiments or representatives of certain spiritual qualities, so the metals, to the alchemists, embodied these qualities. Indeed, the connexion between the heavenly bodies and the metals was clearly defined. Thus gold is the earthly image or representative of the sun. Silver represents the white moon. Venus is represented by copper, Mercury by quicksilver, Mars by iron, Jupiter by tin, and Saturn by lead. In this universe, also, everything is endeavouring to attain a higher state. So that in trying to transmute other metals into gold the alchemists were merely helping on a natural tendency.

In supposing that substances were constituted by their qualities, and that by changing their qualities the substances could be changed, the alchemists were pursuing a perfectly reasonable idea. It is strange, however, that they attached so much importance to the quality of colour. The famous Arabian physician and philosopher, Avicenna, was almost alone in maintaining that the difference between metals was more deep-seated than their colour, and that by the mere changing of colour one could not bring about the transmutation of metals. The belief in the primary importance of colour persisted, however. The famous alchemical doctrine of the three principles, sulphur, mercury, and salt, seems to be connected with this belief. It was found that mercury and sulphur combine to give a brilliant red sulphide, and it was felt that this colour testified to the existence of something even more noble than the white of silver or the yellow of gold. These substances, sulphur and mercury, became, as it were, generalized, and were elevated into principles. Thus sulphur became the principle of fire. It was the principle of sulphur that enabled a body to be combustible, and it was sulphur that disappeared on burning. Mercury, which slips so easily from between the fingers, embodied the principle of liquidity. It was the mercury contained in a burning body that caused part of it to distil over in the liquid form. Salt represented the earthy or solid principle, and was the residue left over when a substance was burned. This outlook persisted until the middle of the seventeenth century.

§ 2

It was Robert Boyle who first showed that a radically different classification of material substances was necessary. In the year 1661 appeared his book entitled *The Sceptical Chymist: or Chymico-Physical Doubts and Paradoxes, touching the Experiments whereby Vulgar Spagirists are wont to Endeavour to Evince their Salt, Sulphur and Mercury to be the True Principles of Things*. In this work he showed, amongst other things, that heat does not always resolve a substance into its elements. Different degrees of heat produce different effects, and sometimes we are left, at the end of the process, with what is obviously a complex substance. He found, also, that gold, after being dissolved in *aqua regis,* could be recovered in its original form, a fact which suggested that there were atoms of gold which persisted unchanged through the transformation. He suggested, therefore, that there are distinct and persistent substances out of which other substances are composed, and thus expressed the cardinal chemical distinction of elements and compounds.

This new classification of material substances was a very great step in advance. The next step was to determine which substances were elements and which compounds, a task which took many years. Success was considerably delayed by the old and persistent puzzle regarding the nature of combustion. When a substance burns, it seems that something escapes and, as we have seen, the alchemists called this something sulphur. It was renamed phlogiston about the beginning of the eighteenth century, and this phlogiston was endowed with the very remarkable property of possessing negative weight. When added to a body it made the body lighter. This was necessary, because Boyle and others had shown that when a metal was burnt the resulting solid was heavier than the original metal. Therefore, since the absence of phlogiston made the metal heavier, phlogiston must have weighed less than nothing.

That air was a material substance, possessing weight, had been shown by Boyle. But it was not known that air was a complex substance, a mixture of gases. Indeed, the only gas recognized to exist was air. Nevertheless, a number of experiments had obtained other gases, but these gases were all regarded as air in a peculiar condition. Thus Joseph Black of Edinburgh discovered, about 1755, carbon dioxide. He named this gas "fixed air." Joseph Priestley actually discovered oxygen, and showed its power in supporting combustion and its necessity for the respiration of animals. But he did not in the least realize what he had done, and called this substance "dephlogisticated air." Priestley was perhaps a rather stupid man, but it is amazing that Henry Cavendish, that extremely acute mind, was also content to accept the phlogiston theory. He disproved the notion that water is an element by analysing it into its constituents, oxygen and hydrogen, but he called these two gases phlogiston and dephlogisticated air. The notion of phlogiston was, indeed, one of the most misleading notions that has ever been introduced into science.

The true theory of combustion was finally reached by the great primary scientific method of weighing and measuring. Lavoisier, the great French chemist, who gives one the impression of being a professional contrasted with amateurs, attacked the problem by this method, and with triumphant success. He heated a known quantity of mercury enclosed in a vessel containing a known quantity of air. A red rust formed on the mercury and the quantity of air diminished. The red rust, on being strongly heated, split up into metallic mercury and a gas. The gas so liberated was found equal in quantity to the shrinkage of the air in the first experiment. Hence nothing had been lost. Lavoisier found that the air left in the first experiment, after the shrinkage had taken place, was incapable of supporting life or combustion. The air released from the mercury rust, on the other hand, supported both to an extraordinary degree. Lavoisier comments: "Reflection on the conditions of this experiment shows that the mercury on calcining absorbs the salubrious and respirable portion of the atmosphere, and that the portion of the air which remains behind is a noxious kind of gas incapable of supporting combustion and respiration. Hence atmospheric air is composed of two elastic fluids of different and so to speak opposite nature." This experiment was the death-blow of the phlogiston theory.

With the phlogiston theory out of the way, systematic advance in the classification of matter became possible. It became recognized, for instance, that one and the same substance can be susceptible of three states, solid, liquid, and gaseous, and this fact, in its turn, led to the recrudescence of the atomic theory. That theory, that all matter consists of little indivisible particles, was very old, as we have seen. It had been known to the Greeks, and it had been regarded favourably by the great founders of the modern scientific movement, Galileo, Newton, Boyle. But now, through the labours of a Manchester schoolmaster, John Dalton, it, acquired a fruitful, that is to say, a quantitative, form. Dalton distinguished clearly between elementary and compound substances. Compound substances are built up out of two or more elementary substances. Dalton knew twenty elementary substances; we now know ninety-two. Dalton established the atomic theory on a quantitative basis by pointing out that the atoms of different substances have different weights, and that elements always unite in the same proportions to produce the same compound.

With the atomic theory as its chief weapon, the analysis of matter, so far as it could be conducted by chemical means, was launched on its modern career. But the further development of chemistry has contributed nothing of the first importance for our understanding of the main lines on which the material universe is constructed. The immense and intricate labours of the chemists have revealed to us, in detail, the structure of an almost unbelievable variety of substances, and the practical importance of this knowledge, both for peace and war, is, as we know, very great indeed. But the region covered by chemistry, although complex, is circumscribed. It is only when it merges into physics that we reach really great new generalizations.

The great contribution of chemistry to our general scientific outlook is its demonstration of the theory that matter is composed of atoms. The atom of the chemists was envisaged as a little hard solid particle, probably spherical in form. This analysis of matter into atoms was considered to be ultimate, although it was certainly felt as vaguely disturbing that there should be ninety-two different kinds of atoms. It was even surmised that atoms had not, perhaps, the simple structure they were supposed to have, for certain groups of elements were found to have marked affinities with certain other groups, suggesting that their atoms were constructed on very much the same plan. It had been suggested, before the technique of atom-weighing was well developed, that all atoms were built up out of hydrogen atoms (hydrogen being the lightest element), and that therefore all matter was really one substance, namely, hydrogen. But more accurate measurements showed that the atomic weights would not fit in with this hypothesis, and therefore chemists had to accept the existence of ninety-two kinds of matter as being another instance of nature's inexplicable diversity.

§ 3

The modern era may be said to have begun about the year 1895. Between that year and the year 1900 those researches were begun which have changed our whole conception of matter and which have, further, changed our whole idea of the meaning and purpose of physical science. So great a revolution in scientific thought has not occurred since Copernicus showed that the earth went round the sun.

The first step was the experimental demonstration

that there exist little electrified bodies very much smaller than a hydrogen atom. These little bodies were produced by passing an electric current through a glass tube almost entirely exhausted of air. It became evident that a stream of something passed in straight lines from the electric plate at one end of the tube to the electric plate at the other end of the tube. The nature of this stream was carefully disentangled by experiment, and it was found that a considerable portion of it consisted of little electrified bodies nearly two thousand times smaller than the known size of a hydrogen atom. There does not, at first sight, appear to be anything very revolutionary about this discovery. Atoms were already unimaginably minute, and to be told that still smaller particles exist comes as no particular shock, although it is a little surprising, perhaps, to find that they are electrified. But a closer investigation of these little particles, or electrons, as they were called, led to a really revolutionary change in our ideas about matter.

It had been known, by mathematical reasoning, that a body charged with electricity behaved as if it had acquired a greater mass than it possessed before being charged. It became a matter of interest, therefore, to determine how much of the mass of an electron was due to its electric charge. The measurement was made, with the truly astounding result that the *whole* mass of the electron was found to be due to its electric charge. The electron was not a piece of "ordinary matter" at all; it was nothing but an electric charge. Now this result is not only interesting and astonishing; it also has a quite peculiar kind of importance. For it was the first indication to come from modern science that the material universe is not the substantial, objective thing we had always taken it to be. Matter began to thin away into the completely spectral thing it has now become. The new discovery could not, of course, be immediately assimilated. To talk of an electric charge existing apart from "ordinary matter" seemed to many minds as incomprehensible as talking about motion without anything that moves. Indeed, the discovery could not be understood until our notions of matter became more abstract. The notion of *substance* had to be replaced by the notion of *behaviour*.

The notion of substance is a very difficult notion to define. A piece of matter may possess the properties of being hard, cold, coloured, and so on, but what is the something, the *substance,* that possesses these qualities? When we have abstracted from the qualities a piece of matter possesses, what have we left? This notion of substance, when we think of it, seems to be associated in some way with our sense of muscular effort. The only scientific, measurable quality of matter that seems to be akin to this is the quality of inertia.

Inertia is that quality of matter, our school-books tell us, in virtue of which a piece of matter will not move from a position of rest until a force acts on it.

GOSHEN COLLEGE LIBRARY
GOSHEN, INDIANA

It is the quality which makes anything "substantial" or "concrete." Light, for instance, used to be regarded as something quite immaterial and unsubstantial. But now that we know that light exerts pressure on any object on which it falls, we are willing to dower it with some degree of substantiality. Our difficulty, then, in admitting that electrons, "disembodied charges of electricity," can play the role of matter, is appreciably diminished when we learn that electrons manifest the property of inertia. In this respect they *behave* like matter. In other respects, also, electrons behave like bits of matter. An electron has, for instance, locality. For all practical purposes an electron exists in a definite region of space. We say "for all practical purposes" since, theoretically, the influence of an electron extends throughout space. But beyond a certain very small distance this influence is imperceptible. An electron can also be tracked through time and space, that is to say, it is capable of motion, and motion in a definite path. This statement requires some modification, as we shall see later, but it is sufficiently true at the present stage of our argument. These characteristics, taken together, enable the electron to play the role of what we have called a material particle in a sufficiently convincing manner.

So far as we have gone, then, no intolerable strain is made on our pictorial imagination. At this stage we can, without being utterly wrong, picture an electron as a little round object. The next step is to form a picture of matter as constituted of electrons. We have to picture the various kinds of atoms as built up out of electrons. In the first place, it is evident that electrons alone are not sufficient to compose an atom. All electrons consist of charges of the same kind of electricity—the so-called "negative" electricity. But atoms are not normally electrified; they are electrically neutral. It is evident, then, that each atom must contain a charge of "positive" electricity sufficient to counterbalance the negative electricity of the electrons it contains.

It took some little time to obtain a satisfactory model of how these charges are distributed in an atom. It was suggested, for instance, that the positive charge formed a kind of sphere in which the electrons were embedded. Certain experiments by Rutherford, however, showed that this idea would not do. It became clear that the positive charge, instead of being diffused throughout the body of an atom, must be regarded as very highly concentrated, and situated in the middle of the atom. The electrons were then pictured as circulating round this central nucleus. This picture is the celebrated "miniature solar system" picture of the atom. On this theory, which is still, in essentials, the accepted theory, an atom is nearly all empty space. The diameter of an electron is about one fifty-thousandth part of the diameter of the whole atom—that is, of the diameter of the orbit in which the outermost electron is revolving. The diameter of the nucleus is about the same as that of an electron. If all the atoms of a man's body were so condensed as to leave no unfilled space, the man's body would become a barely visible speck. According to this theory so-called "solid" matter is an extremely tenuous thing.

The experimental evidence for this theory is extremely good, and one of the most striking illustrations of it is provided by radium. The radium atom is a very heavy atom, and the heavier an atom the more complicated is its structure. In a heavy atom there are several circulating electrons, and also the nucleus itself is a complex affair. The simplest and lightest atom of all, the hydrogen atom, consists of a simple positive charge in the middle with one electron revolving round it. This simple positive charge is called a proton. It carries an electric charge of the same magnitude as the electron, but of the opposite kind. It is, however, very much heavier than an electron. Indeed, practically the whole weight of an atom is due to the protons it contains. In all atoms except hydrogen the nucleus consists of a combination of protons and electrons, with the protons preponderating. In the heaviest atoms this structure is so complex that it seems to be unstable. The nuclei of these atoms are spontaneously disintegrating.

This process is particularly violent in the case of radium. Nothing that we can do, no variations of temperature or pressure, has the slightest effect on the rate of disintegration. The radium atom, in breaking down, shoots out three kinds of rays, which are usually distinguished by the three letters of the Greek alphabet, α, β, γ. The α-rays consist of a stream of particles, each particle consisting of a combination of four protons and two electrons. An α particle is, indeed, identical with the nucleus of a helium atom, the atom which comes next to hydrogen in the scale of complication and weight. The β-particles are simply electrons. The γ-rays are not particles at all, but a sort of very short X-rays. All these rays proceed from the highly complicated nucleus of the radium atom, and in this phenomenon we have a very good confirmation of our general theory of atomic structure. But at present we are quite ignorant of the reasons for this spontaneous break-up.

In this general survey there is no need to go into the experimental verification of the theory in more detail. We may take it that the theory is, in its general lines, fully confirmed. But we must say something about the source of energy, contained in the atom, that this theory reveals to us. The heat given out by an ounce of radium in an hour is sufficient to raise one and a half times its own weight of water to boiling point. This amount of energy is comparatively insignificant, for only a small proportion in any collection of radium atoms are disintegrating at any one moment. By building up the nucleus of a light element, such as helium, out of protons and electrons, we would have a more intense form of energy. By building up an ounce of helium, for example, we would have enough energy to run a one-hundred-horse-power engine continually for eight years. But incomparably the most intense source of energy would be obtained by annihilating matter. If we could make a proton and an electron coalesce, they would both vanish in an intense flash of radiation. The energy that could be obtained in this way is almost inconceivable. The annihilation of a thimbleful of mud, for example, would provide sufficient energy to drive a large ship across the Atlantic.

There is some evidence that this process is actually taking place—not on earth, but in the stars. The stars are continually radiating enormous quantities of energy, and it is a celebrated problem in astronomy to determine where this energy comes from. A very important element in the problem is the age we are to attribute to the stars. Modern estimates give the average age of the stars as some millions of millions of years, and this age, if we accept it, at once rules out certain suggested sources of energy. They could not provide the observed radiation for a sufficient length of time. The most plausible supposition appears to be that the matter of the stars is being annihilated. On this theory the sun, for example, is losing three hundred and sixty thousand million tons of matter every day, but its mass is so enormous that it can go on doing this for another fifteen million million years.

This theory, as we have said, is the one that best fits in with an estimated age for the universe of some millions of millions of years. The new theory of the expanding universe, however, throws considerable doubt on this particular time-scale. It becomes very unlikely that the universe is as old as that. It is much more probable that the age of the universe is to be reckoned in thousands of millions of years instead of millions of millions. It then becomes possible to suppose that the energy of the stars comes from the building up of matter rather than from its annihilation. The building up of hydrogen atoms into helium atoms would provide enough energy to account for the observed radiation on the revised time-scale. If the sun were originally pure hydrogen its transformation into helium would provide energy at the present rate for a period of one hundred thousand million years.

Is the matter of the universe being annihilated or is it being built up, or are both processes going on? It is not yet possible to give definite answers to these questions, and there is considerable diversity of scientific opinion regarding them. About thirty years ago an exceedingly penetrating kind of radiation was discovered traversing the atmosphere. This radiation does not come from the earth, for balloon expeditions showed that it is more penetrating at great heights than at sea-level. Also, it does not come from the sun, for it is not more abundant at day-time than at night-time. The sun is quite an average, typical star, and therefore, as the radiation does not come from the sun, there is no reason to suppose that it comes from the stars. It must come from outer space. What is its origin?

This "cosmic radiation," as it is called, cannot be released by any of the natural processes known on earth. The most penetrating radiation produced by known means are X-rays and the γ-rays from radium. X-rays will penetrate a few millimetres of lead. γ-rays will pass through some inches of lead. The cosmic rays will pass through sixteen feet of lead. If we assume that these rays are analogous in constitution to X-rays and α-rays their wave-length can be calculated from their penetrating power. It can then be shown that waves of this kind would be produced by the annihilation of hydrogen atoms or, more generally, by the coalescence of a proton and an electron, a process that might occur in other atoms than those of hydrogen. Thus these rays would point to the annihilation of matter taking place in remote regions of space. On the other hand it has been pointed out, particularly by Millikan, that the building up of certain very complex atoms would produce rays of this kind. It may be that they testify to a process of "creation" going on in the universe.

This theory, however, appears to be less probable than the annihilation theory, for it is difficult to imagine how these complex atoms could be built up at one blow, as it were, and any gradual building up would not account for the observed radiation. At present the question cannot be decisively settled. It may be, also, that the cosmic rays are not waves analogous to X-rays but particles, analogous to β-particles. In this latter case their speed must be very nearly equal to that of light to account for their observed penetrating power. But this alternative, it is generally accepted, is less probable than the former.

§ 4

So far we have envisaged the electron as being, in all essentials, a little particle. This conception of the electron was very fairly successful up to the year 1925. Since then it has been found that the electron also possesses another group of properties—the properties peculiar to waves. The electron behaves both as a particle and as a group of waves. Certain experiments have been designed in which the electron unmistakably manifests itself as a particle. In other experiments it just as indisputably manifests itself as a group of waves. We can combine these properties in a word —Eddington has suggested the word "wavicle"—but we cannot combine them in a physical picture of the electron. The electron, it seems, is now definitely outside the possibilities of the pictorial imagination.

Nevertheless, we can reason about it and say what ought to happen in certain circumstances. For in order to reason about it we require only to know its mathematical specification. We replace the picturable electron by a collection of symbols. We do not know what these symbols stand for in actual physical reality, but we do know the laws they must obey. We can therefore say what will occur to anything of which these symbols are representative. Twice two is four, whether we are talking about cabbages or kings. The mathematician, therefore, is not distressed by the fact that he cannot picture the electron as a physical entity. Provided that he can construct a coherent mathematical web about it, he knows all about it that he needs to know. But not all men, not even all scientific men, are mathematicians, and this new stage that physical science has reached has taken it distressingly far from ordinary comprehension.

Yet something of the sort was to be expected. Why should we suppose that the fundamental processes of nature should resemble processes made familiar to us by ordinary experience? This assumption was very characteristic of the British physicists of the nineteenth century. Lord Kelvin declared that he could understand nothing of which he could not make a working model. So far, then, as nature does not behave in a way that a nineteenth century engineer could reproduce in his workshop, Kelvin admits that he could not understand it. This singularly arbitrary criterion was not imposed by the Continental physicists. They were content with a much more abstract use of the word "understanding." And it is what we may call the Continental type of mind that is now the most successful in developing physical science. At the present stage of physical inquiry the man who cannot think except in terms of pictures is at a marked disadvantage.

The present state of affairs is perhaps even more disastrous for the writer who wishes to present modern scientific ideas to the layman. For when the layman asks for an explanation he is saying, nine times out of ten, Show me a picture. And when the expositor is dealing with something which is both a group of waves and a particle he is in a quandary. If the men of science are not to become a remote and detached priesthood, scientific ideas must either once more become condensable into familiar images, or else the layman must develop a novel passion for abstract logical relations. We have already had an instance, in the theory of the finite but expanding universe, of how irrelevant, even positively hampering, the pictorial imagination is to these modern intellectual constructions. When the mathematician says that space is finite and expanding he is saying that spatial relations, measured distances, more conveniently fit into a certain mathematical scheme than into any other. And the logical development of this mathematical scheme leads to the conclusion that space is finite and that it is expanding.

Nothing is gained by trying to picture this space. The attempt to do so leads to such questions as "If space is expanding, what is it expanding into?" This question *presupposes* a picture, and is therefore irrelevant to the mathematician. The objection conveyed by this question, if translated into terms to which the mathematical mind is susceptible, would run: "Your theory of a finite but expanding space is logically self-contradictory." The mathematician would certainly pay attention to this objection, and be greatly interested in the argument that professed to establish it.

Seeing that our understanding of nature has now reached a stage when we cannot picture what we are talking about, it need be no shock to us to learn that the laws obeyed by the electron are as queer as its constitution. But when the scientific man talks about the extraordinary character of physical laws, what precisely does he mean? What sort of laws are we to regard as ordinary, and in what way do the laws of modern physics differ from these? The discussion of this question will occupy us in the next section.

CHAPTER III

The Web of Reason

Amongst the phenomena first investigated by man, in his endeavour to find the laws governing natural happenings, were the phenomena of motion. The Greeks, at one time or another, held different views on this matter, but the views that became most popular were all incorrect. They were generalizations based on the most obvious experiences. They held, for instance, that a body only moves as long as there is a force to keep it moving. This is borne out by obvious experience. Thus if we start a stone rolling along a road it will sooner or later come to rest, unless we keep on rolling it. It will go further on a sheet of ice, but even there it will ultimately come to rest. The Greek view, on the evidence, may be called the common-sense view.

The first man to publish correct ideas on motion seems to have been Galileo, although even he did not attain the perfectly clear grasp of the subject that we find in Newton. But Galileo knew that the natural tendency of a moving body is to persist in its motion in a straight line and at a uniform speed. To the Greeks different bodies had different natural motions.

Thus the natural tendency of flame was to fly upward, seeking its natural place in the heavens. The natural tendency of heavy bodies was to sink downwards. The heavenly bodies naturally moved in circles, for the Greeks were greatly impressed by the æsthetic properties of the circle, and regarded it as the most perfect of all figures, and thus peculiarly appropriate to heavenly bodies.

It was the more difficult for the early thinkers to disentangle the idea that the natural motion of a body is motion at a uniform speed in a straight line, as such motion is never observed on earth or, indeed, anywhere else in the universe. As we have already said, even a stone sent skimming across a sheet of ice gets slower and slower and comes to rest. Here the hampering agent is friction. Even a falling stone does not furnish an illustration of natural motion, for it does not fall at a uniform speed. Again, a stone thrown slant-wise into the air does not move in a straight line. It describes a parabola and falls to the earth again. In these cases the hampering agent is gravitation. The "natural motion" of a body, free from the influence of all forces, is something of which we have no experience. For, if all other forces be removed, gravitation remains; it is omnipresent throughout the universe. Why, then, did scientific men choose, as the foundation for their reasoning about motion, a law that can never be verified by observation?

They chose the law because it was the most convenient possible law to choose. It introduced an unrivalled simplicity and economy into the complicated phenomena of motion. For it must be remembered that what scientific men mean by truth is, in the last resort, convenience. Scientific men are pragmatists in practice, whatever they may think they are in theory.

The first law of motion is an example of this. It describes an unobservable state of affairs, but all observable states of affairs can be accounted for much more simply if we assume it. It represents what scientific men call a "limiting case," something we never reach, but which we can approach closer and closer. If we make our sheet of ice smoother, for example, our skimming stone will move more nearly in accordance with the first law of motion than it did before. We notice, in the heavens, that bodies more nearly obey the first law of motion the farther they are from the gravitating influence of other bodies. The first law describes what would happen if there were no disturbing forces, and the fact that what it describes never does happen is explained by the fact that there always are disturbing forces. This procedure is perfectly logical although Einstein, as we shall see later, has suggested a different one.

The phenomena of motion, although pretty generally regarded, both in ancient and in modern times, as a rather dull subject, is nevertheless of immense importance for the understanding of nature, and it is astonishing that correct ideas on it should have been so late in coming to birth. Perhaps it was boredom with the subject, a feeling that it was trivial, that prevented men from giving it closer attention. However, it is certainly surprising to find that, even in the seventeenth century, men believed that heavy stones fall faster than lighter ones, and that the difference in speed is proportional to their weights.

Galileo demonstrated the falsity of this view by his celebrated experiment of dropping different-sized round shot from the top of the leaning tower of Pisa. This is surely one of the simplest experiments recorded in the history of science, and it is difficult not to believe that it must have occurred to millions of men before him. But apparently it did not, and one can only suppose that men were too uninterested in the problem of falling bodies even to pick two differently sized stones from the ground. But Galileo's further experiments on falling bodies were a great deal less obvious. He knew, like everybody else, that a stone, in its fall towards the ground, falls faster and faster. But he claims to have been the first man to ask himself *at what rate* the speed increased. This problem, he tells us, had never occurred to any other philosopher, although many of them had written enormous treatises on the subject. Galileo, by simple but ingenious experiments, found out the law of increase. By these and other experiments he was the first to put the study of motion on a really scientific basis. And Newton, born the same year that Galileo died, showed that in this study lay the key to the material universe. Thus the study which, so far as we can see, must have seemed to the majority of human beings the most boring of all pursuits, turned out to be the initiation of a new epoch in human thought. But before we describe Newton's establishment of the reign of law we must say something about Kepler's researches on the motions of the heavenly bodies.

Kepler had one of the most wildly luxuriant imaginations that any man of science has ever possessed. He was extraordinarily fertile in hypotheses, and many of them, from our modern point of view, were extremely absurd. But he had the great saving grace of being docile in face of the facts. However attractive a theory appeared to him, he always put it to the test of observation and calculation. Thus when he had made what we now regard as the entirely puerile discovery that the "five regular solids" could be inserted between the orbits of the six planets known to him, he says:

> The intense pleasure I have received from this discovery can never be told in words. I regretted no more the time wasted; I tired of no labour; I shunned no toil of reckoning, days and nights spent in calculation, until I could see whether my hypothesis would agree with the orbits of Copernicus or whether my joy was to vanish into air.

Kepler did, indeed, attach just as much importance to the discoveries we no longer value as to those we regard as an integral part of science. The research we have just mentioned, for example, was undertaken to find out why God had made exactly six planets. Kepler was a remarkable case of the man who possesses immense scientific abilities without possessing the scientific outlook. But this was not, perhaps, so very remarkable at his time, for the scientific outlook was not yet fully conscious. It first became fully conscious in the person of Isaac Newton.

The laws with which Kepler's name is for ever associated are the three laws of planetary motion. Copernicus, although he had broken away from the centuries-old Ptolemaic system, and declared that it was the earth, and not the sun, that moved, was still impressed by the principle of circular motion. He taught that the earth and the other planets describe circles round the sun. Kepler was also inclined to accept this principle, but his admiration for the æsthetic charms of the circle could not prevail against his fidelity to the facts. He made eighteen successive attempts to fit his observations (or rather Tycho Brahe's, that queer observational genius from whom Kepler derived them) to circular orbits, and only then did he throw the principle of circular motion overboard, and declare that the planets moved in ellipses.

Nevertheless, although Kepler had parted with the principle of circular motion, he felt that he had preserved its essence when he went on to discover his second law. The great point of circular motion was that it gave a dignified uniformity to the motion of the planets —a uniformity that was felt to be appropriate to heavenly bodies. Kepler found that this uniformity was still, in a way, preserved. For although a planet, by moving in an ellipse, changes its distance from the sun, and also changes its speed, there is an aspect of the motion that remains constant. If we imagine a straight line drawn from the sun to the planet, then, as the planet moves, this line will sweep out an area. Kepler discovered that this line sweeps out equal areas in equal times. As the planet, in describing its elliptical orbit, approaches closer to the sun, its speed increases, and in such a way that the decreased length of the line from the sun is compensated for by the greater speed, so that the area it passes over grows at a perfectly uniform rate. Kepler saw in this law some sort of substitute for the simple nobility of the principle of circular motion. However that may be, it is a testimony to his scientific instinct that he isolated just the element which is uniform and constant in this complicated motion.

Kepler's first two laws are true for each planet individually. It is true of each planet that it moves in an ellipse, and that its radius vector sweeps out equal areas in equal times. Kepler's third law finds a relation between all the various planets. The planets are situated at different distances from the sun, and the time they take to describe their orbits varies. The outer planets, those farthest from the sun, have not only farther to go, but they move more slowly. Kepler found a law showing how these relative times are connected with these relative distances. With Kepler's three laws we have, for the first time, a complete description of the planetary motions.

But these laws are purely descriptive laws. They are just statements of fact, like saying that gold is yellow. Kepler gives no reason why these laws should be as they are. The observation and recording of laws is the first step in scientific procedure. Science begins by hunting for uniformities amongst natural phenomena. The scientific man finds, for instance, that light is propagated in straight lines, that unsupported stones fall to the ground, that heat passes from a hotter body to a cooler one. In this way he introduces a sort of order into whole groups of happenings. And this sort of knowledge is often quite sufficient for practical purposes. Indeed, in many cases of great practical importance science has not yet advanced beyond this knowledge.

But the aim of science is to proceed to the next step, the construction of theories. Just as a law unifies a group of phenomena, so a theory unifies a group of laws. A theory is, as it were, a central principle, from which the various laws belonging to it can be deduced. The outstanding example of this is provided by the most celebrated of all scientific theories, Newton's theory of gravitation.

Newton knew many isolated laws. He knew Kepler's three laws of planetary motion, he knew the rate at which a falling body increases its speed, he knew that the path of a projectile is a parabola, and so on. He saw that if we suppose that every piece of matter in the universe attracts every other piece of matter we can show that all these diverse phenomena are necessary consequences of this attraction. The strength of the attraction depends, of course, on the masses of the bodies involved, and also on their distance apart. Newton gave the precise formula according to which the attraction varies. From his formula he deduced, amongst other things, Kepler's three laws of planetary motion. He showed that they were not disconnected; they all flowed from this one general principle. Complete order and harmony were established throughout the solar system by Newton's theory and, as later astronomers were to show, throughout the heavens.

It was the first demonstration, on a grand scale, of the mathematical machine aspect of Nature's processes, and inspired scientific men to extend this reign of law to cover all natural phenomena. The attempt has been extremely successful, but certain important elements in the general outlook on which it was based have, in our own day, been abandoned. But before we can understand this modern revolution we must look a little more closely at the nature of Newton's achievement.

§ 2

Science, in its attempt to understand the world, has inevitably had to proceed by fashioning its images and concepts out of ordinary experience. A striking example of this kind of anthropomorphism is the concept of force. When Newton says that two bodies *attract* one another, what does he mean? Are we to picture the two bodies as pulling at one another? If so, we must imagine the pull being somehow conveyed across the intervening space. Otherwise, if we have one body influencing another without being in any sort of physical contact with it, then we are either in the realm of magic or in the realm of psychology. We know that two people who "attract" one another may "will" to come together, but we can hardly imagine such a relation existing between the sun and the planets.

But although Newton regarded "direct action at a distance" as so great an absurdity that, as he says, no man with a competent faculty for thinking could possibly fall into it, yet his theory assumes it. Newton's formula for gravitation takes no account of time. The influence, whatever it may be, is supposed to act instantaneously. But Newton was not dogmatic about this. He speculated as to whether there was not some subtle medium which conveyed the influence. He said that his formula merely described the way in which bodies behaved; it did not profess to say anything about the causes of that behaviour. Indeed, he did not think that science could say anything about ultimate causes, for ultimately, Newton thought, everything happens by the will of God.

But the mathematicians after Newton, particularly the Continental mathematicians, erected direct action at a distance into a fundamental principle. They dowered matter with a variety of attractions, electric, magnetic, gravitational, and a few repulsions. All these acted instantaneously at a distance. The result was that the intervening space, in which we now know that very important physical processes are going on, was utterly ignored. This was, of course, quite natural on the assumption that these effects are propagated instantaneously. Nevertheless, the mathematicians imagined that the whole of space was filled with a vast jelly-like substance called the ether. But the sole function of the ether was to account for the propagation of light which, it was known, travels at a finite velocity.

The nature of light is still one of the fundamental puzzles of science. The only period in scientific history when men of science thought they knew all about it was during the first half of the nineteenth century, when the ether theory was being worked out. Since then the nature of light has become what it was before, a mystery. There have always, of course, been theories about the nature of light, but in early times, as in our own, they do not seem to have been held with any great degree of confidence. Certain of the Greeks held the singular idea that the visibility of objects is created by the eye. Thus Epicurus thought that sight is caused by the projection of pictures from the eye. Plato modified this theory by supposing that there is a meeting of rays, some of which emanate from the eye and some from the visible object. And long after this queer theory had been abandoned it was still thought that light was propagated instantaneously.

This is, indeed, a very natural assumption to make. The idea that light is something that moves with a definite velocity is, when one comes to think of it, a very queer idea. That we should *see* when we open our eyes seems the most natural thing in the world. It seems strange to apply to light, this circumambient something, the notions of distance and velocity. The notion that light moves must have been one of those amazing and unexpected leaps of the imagination in which the chief charm of science will always reside.

The first man to put forth this idea and to furnish proofs for it, was Römer, a Danish astronomer. In the year 1675 he was watching the movements of the first satellite of the planet Jupiter. As the satellite passed round the back of the planet it was, of course, eclipsed. Römer measured the time between the satellite's disappearance and re-emergence. But he found, as he continued his observations, that this time varied. Jupiter and the earth, in their revolutions round the sun, are sometimes approaching one another and sometimes receding from one another. Römer noticed that the variations in the satellite's period are connected with this fact. The brilliant idea occurred to him that these variations could be explained if we suppose that light takes time to traverse the distance from Jupiter to the earth. For in one case the earth is advancing to meet the light, and in the other case it is retreating from it. With the instruments at his disposal Römer could not make very accurate observations, and the value he deduced for the velocity of light is too high. We now know, by refined terrestrial experiments, that light moves with the velocity of 186,000 miles per second.

The fact that light takes time to travel makes the question of its propagation acute. Newton favoured the idea that it consists of little particles shot out from the visible object, but this theory, as he expounded it, was by no means a straightforward theory. It was, in certain respects, singularly like the modern theory, which asserts that light is a queer blend of waves and particles.

Newton's influence served to discredit the wave theory during his time, but by the beginning of the nineteenth century it was flourishing in full vigour. The wave theory of light, which is really the ether theory, is one of the most amazing instances in science of a theory which is developed with enthusiasm by extremely gifted men, which goes on from success to success until it secures the universal assent of the scientific world, and yet which turns out to be wrong. An even more startling instance is provided by Newton's theory of gravitation, which was regarded for over two hundred years as "the most perfect, and perfectly established, of scientific laws." Its accuracy and perfection became proverbial; it was the ideal by which all other scientific theories were judged. And great mathematicians have been unanimous that Newton's discovery and development of it was the supreme achievement of human genius. Yet today we account for gravitation in entirely different terms. Our whole outlook on this question is radically different from the Newtonian outlook. It is not only that the Newtonian theory leads to results which are incorrect in detail; the whole method of thought on which it reposes has been abandoned, root and branch.

This fact need not give rise to melancholy reflections. Science is a wonderful exhibition of stages in the greatest of human adventures, the intellectual adventure. In the refinement and subtilizing of his emotions and bodily sensations it is doubtful if man has made much of an advance. It is doubtful whether, in these matters, we have advanced much beyond the Greeks and the Romans, or whether we know more in these matters than was known to the great civilizations of India and China. But there has certainly been a real and advancing exploration of the resources of the mind. Man has advanced from one level of abstraction to another. He thinks thoughts that were never thinkable before; he has broken through mental barriers that never before were broken. Whatever may be true of the physical universe, it is true that the mental universe is constantly expanding. Newton's theory was the most perfect that could be constructed in the universe of ideas in which he lived. And his curious indifference to science, and his passionate interest in mysticism, was probably due to the awareness of the fact that his universe of ideas, in which he was so triumphant a master, was limited and insufficient. He would probably feel the same today. Genius of so supreme an order is probably always obscurely aware of possibilities of thought that it finds it impossible to formulate. But for the ordinary student the chief charm of science is that it acquaints him not only with a new body of ideas, but with new modes of thinking. He experiences a veritable growth of consciousness.

§ 3

One of the great stages in the evolution of scientific thought was reached in recent years, when the mechanistic explanation of the universe was definitely abandoned. In the Newtonian outlook the notions of matter, space, and time were fundamental. All these notions are now regarded as derivative from a deeper, underlying reality, as we shall see later. The Newtonian outlook is still the outlook of common sense, however. It seems quite natural to think of space as having a being of its own. We can imagine, or we think we can, completely empty space. The fact that space has material bodies existing in it is, it appears, quite irrelevant to the nature of space. We imagine space being spread out before the material universe appeared, and we imagine it continuing to exist even if the material universe were annihilated.

Similarly with time. We *measure* time by events, it

is true. We watch the successive swings of a pendulum or the movements of the heavenly bodies. But these events, it seems, occupy time very much as a body occupies space. Time, in its own nature, seems to be independent of events. We can imagine the march of time going on even in an empty universe in which no events ever happened. But can we imagine time going on in the absence of all space? If there were no space, if the whole universe shrank to a point, would time still exist? This is not so easy to answer. We have an uneasy feeling that the question is meaningless. We become obscurely aware that space and time are, in some way, interconnected. Nevertheless, for our ordinary thinking we assume, as Newton did, that space and time are independent, fundamental realities.

A third reality is matter. It is not a completely independent reality, it is true. We cannot conceive of a piece of matter that does not occupy space or that exists for no time at all. Both time and space are essential to the existence of matter. But matter is a separate reality in the sense that it is, in its own nature, something utterly different from space and time. It is an extraneous something embedded, as it were, in space and time. Out of these three entities scientific men have, until recent times, tried to construct the physical universe. The universe has been regarded as a world of matter moving in an independent space and an independent time.

There were also, of course, a number of subsidiary

concepts. There were attractive and repulsive forces, stresses and strains, cohesion, gravitation, etc. With this particular set of ideas scientific men bound themselves, as it were, to explain—everything. All explanations, of course, proceed within a certain framework of ideas. Thus the psycho-analysts explain all human psychology in terms of their complexes, repressions, etc. The French cynic indicated his explanation of many human misfortunes by saying *"Cherchez la femme."* There are bishops who explain degeneracy, immorality, madness, and crime as due to cocktail drinking or the wearing of short skirts.

The aim of the scientific scheme, also, is to reduce everything to a comparatively small number of fundamental ideas. This is, of course, inevitable, but it has its dangers. The ether theory is a good example of the danger involved in trying to fit reality to an inadequate framework of ideas. The ether theory was an attempt to explain the phenomena of light mechanically. A vast medium was supposed to fill all space, and the mathematicians set themselves the task of working out what properties this medium must possess in order that its vibrations should constitute light. The task proved to be one of enormous difficulty. As the mathematicians piled ingenuity on ingenuity the ether became more and more fantastic. All sorts of strange physical properties were attributed to it. Sometimes, indeed, these properties were entirely contradictory of one another. In the end the ether became more like the nightmare dream of a mad engineer than a physical reality. Space became filled with an incredible complex of gyroscopes, driving bands, pulleys, cog-wheels, and what not.

The construction of ethers had already become less popular (chiefly, one may suppose, through boredom) when Einstein showed that there was no reason to suppose that an ether existed. The whole of this effort to explain light had been a misdirection of energy. Light is not to be explained in mechanical terms.

Up to the beginning of the present century physical science sought for an explanation of the universe in mechanical terms. Even the phenomena of life and of mind, it was hoped, would be brought within this general scheme. The great change that has come over physical science is due precisely to the realization that this particular group of ideas is inadequate. We are in process of replacing them by a different set. But it is not yet clear, unfortunately, what the new set will be. The new concepts, those that we have elaborated so far, are not yet perfectly definite. The electron, for instance, is at present a very hazy entity. It is nothing like so definite a thing as the hard, substantial little atom of the Victorians. And, besides this deplorable haziness, we are not yet sure that we have found the right concepts out of which to build our picture of reality. Our notions of space and time, for instance, although much more subtle and flexible than they were before the appearance of Relativity Theory, are very likely not yet subtle enough. It is suspected that many of the difficulties and puzzles we meet with in science today are due simply to the way we think about things. It is very probable that some of the questions we are putting to nature are just meaningless. It is this fact that makes the present stage of science so peculiarly interesting. We are living, not merely in an epoch of fresh discoveries, but at the birth of a new world outlook.

§ 4

The first definite indication that the old scientific outlook was inadequate may be dated 1905. That was the year in which Einstein published his first paper on relativity. The ether theory, as we have said, seemed to be dead, or at least moribund, under the immense weight of its own ingenious details. It raised more problems than it solved, but it so happens that one of the problems it raised must perhaps be counted as the most valuable of all its achievements. This was the problem of whether the earth, in its motion through the ether, carried any of the ether along with it.

This problem, as we now know, was a pseudo-problem (as are probably some of our problems today), but at the time it appeared perfectly reasonable. Those who believed that the ether was not affected by the motion of material bodies through it maintained that the ether passed through the spaces between the atoms of matter like the wind through a grove of trees. However, the question could not be decided by pure theory, and it was put to the test of experiment. The results were baffling, for some experiments showed clearly that moving matter does not carry the ether along with it, and other experiments showed just as indisputably that it does.

The most famous of these experiments is the one originally carried out by Michelson and Morley in 1887. Since then this experiment has been repeated many times. In principle it is very simple, and consists in comparing the velocity of light in different directions. If the earth is moving through a stationary ether it can be shown that two rays of light, one moving in the direction of the earth's motion, and the other at right angles to it, should take unequal times to cover the same distance. But although the experiment has often been repeated, no difference has ever been found, although in some of these experiments the apparatus has been so delicate that a difference one hundred times less than the difference expected could have been measured. Michelson concluded, from the negative result of his experiment, that the earth carries the ether along with it. This conclusion, however, is quite opposed to certain very trustworthy astronomical observations.

The dilemma thus created is a very real one, and the way out, which was shown by Einstein in 1907, is an effort of genius of the highest order. It is difficult to say which one admires the more, the confidence and boldness necessary to put it forward or the originality and imagination necessary to conceive it. Briefly, Einstein asserted that the velocity of light is always the same, whether we measure this velocity from a system which is in motion or a system that is at rest.

To see the paradoxical nature of this assertion, let us apply it to our ordinary experience. If we say that the speed of a motor-car is sixty miles per hour we mean that its speed, measured by a stationary observer, is sixty miles per hour. If it passes a car travelling in the same direction at forty miles per hour it obviously passes it at a speed of twenty miles per hour. And if the second car, instead of travelling in the same direction, were coming to meet it, they would obviously pass one another at a speed of one hundred miles per hour. Yet, according to Einstein, the speed of a ray of light, measured by these various observers, would be found to have exactly the same value, namely, 186,000 miles per second. The speed of the observer, in fact, makes no difference at all. This is true whether he is advancing to meet the ray of light or whether he is running away from it.

The statement, at first sight, seems to be mere nonsense. It seems to be self-contradictory. But we see a way in which the statement may make sense if we reflect that a velocity is a distance divided by a time. To find the speed of anything we measure the distance it has traversed and divide it by the time it has taken to do it. If, therefore, these different observers are using different measures of distance and of time, it is quite possible that they may get the same ratio. If their clocks are keeping different times, and if their yard measures are of different lengths, then it may happen that these differences balance out, as it were, so that they reach the same figure for the velocity of light.

Einstein asserts that this is what does happen. He says that our measures of length and of time vary with our motion. A stationary observer, watching our measuring instruments as we moved past him, would judge that our yardsticks had contracted, and that our clocks were running slow. Curiously enough, if we watched his measuring instruments we would make exactly the same judgment about them. These extraordinary phenomena have never, of course, been observed, because they do not become measurable, even by the finest observation, except at enormous speeds, speeds comparable with the velocity of light itself.

In order to appreciate the changed complexion given to scientific theorizing by Einstein, we must remember that he starts with an observed fact. That fact is the constancy of the velocity of light. If we admit this as being universally true, then it can be deduced, as a necessary consequence, that space and time measurements are relative in the way shown by Einstein. They vary with the motion of the observer. Two events which are simultaneous for one observer, for instance, are not simultaneous for an observer who is moving with a different motion. There is no such thing as *the* time or *the* distance between two events. Different observers reach different results. And no one observer is more privileged than any other.

We have seen that the Michelson-Morley experiment failed to reveal any motion of the earth through the ether. And as the ether was supposed to be a great jelly filling all space, scientific men would have been quite willing to take motion though the ether as a revelation of motion in absolute space. But no such motion can be detected. Every motion that we can detect is a relative motion. We have to ask, Compared with what is it moving? The standard of comparison chosen is always merely a matter of convenience. It is appropriate to the purpose we have in mind at the moment, but it has no absolute validity. In estimating the speed of a motor-car we take no account of the fact that the earth is rotating on its axis, that it is also revolving round the sun, and that the whole solar system is moving towards the star Vega. Even if we did take these motions into account we should be no nearer knowing the car's velocity "in space." For Vega is taking part in the general rotation of the whole galactic system, and this system itself is moving relatively to others. However far we go we find nothing but a relative motion. The notion of "absolute velocity" is, in truth, unmeaning.

Einstein's theory, as we have seen, explains the null result of the Michelson-Morley experiment. But that is only an incidental result; the theory does a great deal more than that. It explains, as does every great scientific theory, a large number of apparently disconnected facts. And this, of course, gives it a great æsthetic charm. The æsthetic charm of Einstein's theory is, however, quite unusual. We are now speaking of the theory published in 1905—the so-called Special Theory of Relativity. The General Theory, published ten years later, is even more remarkable, and we shall discuss that later.

Einstein has professed to be uninterested in considerations of "elegance" where a scientific theory is concerned. He agrees with Boltzman, he says, that such considerations should be left to the cobbler and tailor. The history of his own theory, however, shows that æsthetic charm and truth are inextricably linked. The true significance of Einstein's theory was not realized until Minkowski gave his celebrated address on Space and Time in 1908. Minkowski gave an astonishing beauty and symmetry to Einstein's doctrine of the relativity of space and time by showing that all this followed naturally if we regard space and time as being aspects of a more fundamental reality—the four-dimensional space-time continuum.

Space, as we have said before in this book, has three dimensions. Time has only one. It is analogous to the one dimension of length. If we wish to know how much of time a man's life has occupied, we ask about its "length." There is nothing corresponding to breadth and thickness in time. The three dimensions of space and the one dimension of time, therefore, if they were in some way amalgamated, would form a total possessing four dimensions. This is the amalgamation that Minkowski announces in his address. As he says:

> Henceforth space by itself, and time by itself, are doomed to fade away into mere shadows, and only a kind of union of the two will preserve an independent reality.

Nature, it appears, knows nothing of the distinction we make between space and time. The distinction we make is, ultimately, a psychological peculiarity of ours. There is nothing absolute about space or time. As we have seen, different observers make different estimates of the space and of the time, separating two events. But there is a certain relation that they are all agreed on—a relation referring directly to the four-dimensional reality. This relation is called the *interval.* It is a spatio-temporal relation, and is obtained by each observer combining his space and time measurements in a certain way. Different observers will disagree as to the space separating two events. They will disagree as to the time separating two events. But they will all agree on the interval between these events.

The significance of this quantity, the interval, was admirably brought out by Minkowski. He showed that, in its mathematical expression, it is analogous to the mathematical expression for a *length.* But, if it be regarded as a length, it must be as a length in a four-dimensional continuum. All observers agree on measurements that refer directly to the four-dimensional continuum.

Now we have already seen that there are various kinds of geometry, Euclidean and non-Euclidean. Each geometry has its own particular mathematical expression for a length. No two geometries are alike in this respect, and from this one expression the mathematician can deduce all the characteristics of a geometry. Now if we regard the "interval" as a length in a four-dimensional continuum we can, by studying its mathematical expression, find what sort of geometry it indicates for the four-dimensional continuum.

The result is very interesting. We find that the geometry in question is not Euclidean, and one of the most striking facts about it is that, according to this geometry, a *critical velocity* must exist. There must be a velocity that appears the same to all observers, whatever their relative motions may be. The fact that we have already discovered such a velocity in the velocity of light is a strong indication, therefore, that we are on the right track. We may regard the Michelson-Morley experiment, in fact, as an experimental confirmation of Minkowski's geometry. Other results that can be deduced from this geometry are also confirmed by experiment. We see that what would seem to us to be singular and arbitrary laws of nature are really consequences of the fact that the four-dimensional space-time continuum is governed by a certain type of geometry. We replace laws of nature by geometry, as it were.

This fact throws a strange light on the scientific picture of the universe. We see that that picture is much more of a mental creation than we had supposed. The attitude of scientific men, before relativity theory, was comparatively naïve. They looked abroad on the universe and saw lumps of matter moving about in accordance with certain laws. Both the lumps of matter and the laws were supposed to exist quite independently of our minds. We simply saw what happened to be there. The lumps of matter we perceived "directly." The laws of matter were not so directly perceived—it was often a matter of considerable trouble to find them—but they were just as truly objective, they were just as truly independent of us.

The attitude today is rather different. We can illustrate it by an example originally given by Poincaré. Suppose we select three heavenly bodies and study the triangle formed by them. According to Euclid the three angles of this triangle add up to two right angles. We make the measurements, we will suppose, and we find that they do not add up to two right angles. Shall we conclude from this that Euclid's geometry is not applicable to our space? We know now that there are non-Euclidean geometries that will explain our measurements. Shall we adopt one of them? It does not follow that we shall. For we have assumed, in making these measurements, that light is propagated in straight lines. We have accepted this as one of the "laws of nature." But this law has been based on a limited number of observations. We might be able to alter the law slightly and still fit these observations. And this modified law, when applied to the great spaces of our astronomical measurements, might give results differing perceptibly from the old law. It might, in fact, account for our measurements, and make it unnecessary to suppose that Euclid's geometry has been violated.

This example suffices to show that the geometry we assume and the laws of nature we discover are not completely independent of one another. We have two variables to play with, as it were. It is all a matter of convenience. We wish to construct our picture of nature as economically and harmoniously as possible. It is obviously desirable to use as few special laws of nature as possible. In the above example we could preserve Euclid's geometry at the price of inventing one special law of nature. In this case we would probably think it worth while to preserve Euclid's geometry. But if we found that the adoption of some non-Euclidean geometry effected a very considerable saving in our laws of nature, then we might find it worth while to adopt the non-Euclidean geometry. This is what Einstein has done in his Generalized Theory of Relativity.

In this theory Einstein has shown that a great deal of nature's behaviour can be explained if we suppose that events are taking place in a non-Euclidean universe. Many of the happenings that have led us to invent laws of nature to account for them are merely natural consequences of the fact that we live in a universe whose geometry is non-Euclidean. As we have seen, Einstein had already done this, to a limited extent, in his Special Theory of Relativity. The General Theory, however, goes very much further, and the geometry it assumes is much less like Euclid's than was the geometry put forth by Minkowski. The geometry adopted by Einstein is, indeed, a very general form of geometry called Riemannian geometry, from Riemann, who invented it. It is a vastly more complex geometry than the one assumed by Minkowski.

Nevertheless, there is one important respect in which Minkowski's contribution was essential to the creation of the generalized theory. That was his discovery that space and time are not to be considered as independent realities. Minkowski's conception of the universe as four-dimensional lies at the very basis of the new theory. As Einstein says, but for this discovery the theory of relativity might never have emerged from its long clothes. But the geometry of this four-dimensional amalgam of space and time is not the comparatively simple geometry assumed by Minkowski. That geometry, as we have seen, accounted for a fair number of things, but Riemannian geometry accounts for many more.

It accounts, for instance, for gravitation. Indeed, this is perhaps its greatest achievement. The force of gravitation had always been a puzzle. In several respects it stands quite alone amongst all the forces of nature.

The gravitational force between two bodies is not affected in the least, for example, by their physical or chemical conditions. Also, there is the puzzling fact that it seems to act instantaneously. Light, as we know, takes time to travel. So does every other form of radiant energy. We have seen that the mathematicians after Newton supposed that the electric and magnetic forces acted directly at a distance, as gravitation was supposed to do. But Faraday and Maxwell showed that the electric and magnetic forces are, in fact, propagated through the intervening space. Maxwell showed that this propagation is in the form of waves, and that these waves move with the velocity of light. This is the discovery, we may remark incidentally, on which "wireless" is based. But all efforts to bring gravitation into line with the other forces proved ineffectual. Many such attempts were made, but none of them could be made to square with observation. It seemed that the instantaneous action of gravitation must be accepted.

Another singular fact about gravitation is that nothing acts as a screen to it. We have substances that stop light, that stop heat, that stop the electric and magnetic forces, that stop even X-rays, but we know of nothing that stops gravitation. A body held up in the air weighs just as much however many bodies we interpose between it and the surface of the earth. The pull of the earth on it is not affected in the slightest.[1]

---

[1] Some of the effects of a screen for gravitation are ingeniously worked out by H. G. Wells in his *First Men in the Moon*.

All these facts are very puzzling, if we conceive gravitation as a force, like the other natural forces. But Einstein showed that it is not a force at all. That bodies "gravitate" towards one another is a straightforward consequence of the fact that we live in a non-Euclidean universe. A four-dimensional continuum governed by Riemannian geometry will possess a certain kind of curvature in the neighbourhood of matter. This curvature manifests itself to us as "gravitation." We can thus understand its omnipresence and unalterability, for it is inherent in the very structure of space-time.

When the formula for gravitation is worked out on this basis it is found to be not quite the same as the formula Newton gave. In nearly every case, however, it is exceedingly close to it, and it is for that reason that Newton's law has been found perfectly satisfactory for over two centuries. As a matter of fact, Newton's law is still used in practice, except in a few very special cases. But these special cases are important, and furnish a beautiful vindication of the truth of Einstein's theory.

One of them is peculiarly interesting, in that it settles a long-standing difficulty. This difficulty is connected with the motion of the planet Mercury, the planet nearest to the sun. On the Newtonian theory Mercury would describe a perfect ellipse round the sun, if it were the only planet in the solar system. But the gravitational influence of the other planets disturbs it, and its actual path is not accurately an ellipse. Astronomers have calculated what this alteration should amount to. But when they compare their calculations with the observed motion of Mercury they find a quite noticeable discrepancy. This fact bothered mathematicians for many years. A famous attempt to solve it was made by Leverrier, the great French mathematician. Leverrier had won great fame by his mathematical discovery of the planet Neptune. He had had nothing to go on but minute irregularities in the motion of Uranus, the outermost planet then known. But from these irregularities, by magnificent mathematical skill, he deduced the size, distance, velocity, and position of the then unknown planet Neptune, and when Dr. Galle of Berlin turned his telescope on the spot that Leverrier indicated to him, he found the planet.

Leverrier, naturally enough, tried a similar method to account for the irregularity in the motion of Mercury. He deduced that there must be an inner planet, rotating between Mercury and the sun. This planet was actually named. It was called Vulcan. Several people thought they saw it, but it turned out that they had only seen spots on the sun. Gradually belief in Vulcan's existence died away, and the puzzle of Mercury's motion remained as formidable as ever. Einstein's theory clears up the whole matter. His formula, when applied to Mercury, gives exactly the motion that is observed. This is obviously a great triumph for Einstein's theory.

An even more spectacular result is Einstein's prediction of the deflection of light. We have already said that Einstein makes gravitation an inherent property of space-time. The planets, for instance, move in the way they do because they are pursuing the easiest path through the space-time continuum that surrounds them. It is not that they are acted on by a force; the path they follow is the natural "straight-line" path in a region governed by that particular kind of non-Euclidean geometry. Now the path of a ray of light must, of course, be just as obedient to the laws of geometry as is any other path. Accordingly a ray of light passing through what we call a gravitational field must follow the path appropriate to that field. This means, in practice, that a ray of light from a distant star, passing near the sun, will be deflected, and the star will appear to be displaced in the sky. This statement can only be checked by observation during times of eclipse, for normally stars in the line of sight of the sun are not visible. As the result of several eclipse expeditions it has been verified that in these conditions, the stars are displaced to the precise extent indicated by Einstein's theory. A third prediction is to the effect that atoms on the sun will vibrate more slowly than similar atoms on the earth. This will influence the quality of the light emitted by these atoms. A spectroscopic examination of the light from the sun confirms this prediction also.

These examples suffice to show that the experimental evidence for Einstein's theory is very good indeed.

Another consideration, which appeals greatly to mathematicians, is the extraordinary inner harmony and elegance of the theory. The results follow so naturally from the premises, and the premises are in themselves so acceptable, that it seems most unlikely that so coherent an argument could be wrong. The later developments of the theory, however, are not so convincing. The ground covered by Einstein's generalized theory is very considerable. It does not, however, account for everything. In particular, electric and magnetic phenomena are left outside his scheme. The later developments endeavour to bring them into the scheme. These developments are due to Weyl, Eddington, and Einstein himself. In each case the procedure is the same. Each of these developments is an effort to account for the electromagnetic forces in terms of geometry, just as Einstein has accounted for gravitation in terms of geometry. The geometry postulated, therefore, is different from the original Riemannian geometry assumed by Einstein, since it can be definitely shown that there is no room, as it were, within that geometry for the electromagnetic forces. Unfortunately, the geometry sought for can be arrived at in different ways or, rather, there is more than one geometry that will do what is required. Our preference for one or the other is, as Einstein has admitted, largely a matter of taste.

The most interesting of these attempts, from the point of view of the general ideas involved, is that of Eddington. For Eddington starts with what would appear to be the minimum number of assumptions possible. He starts with a four-dimensional continuum, and he attributes to this continuum a certain geometry. But this geometry, compared with the firm lines of Euclid's geometry, is so formless, that it may be regarded as but one step removed from complete chaos. Yet from this bare minimum of material Eddington is able, by mathematical analysis, to show that there is something in this welter that has the quality of permanence and that obeys the laws of matter—such laws as gravitation, conservation, etc. Also, the laws of electromagnetism are seen to arise from this primitive structure. It is as if the mind, faced with this inchoate four-dimensional world, selected certain characteristics of it as alone having significance. As a consequence of this selective action the physical world arises. Quite other worlds are also implicit, as it were, in the four-dimensional continuum, and it may be that they would be singled out in preference to our world by minds of a radically different constitution. This feat of Eddington's is extremely interesting, but there are great divergencies of opinion as to its significance. Einstein has said that he does not like the method of approach although, as he admits, he cannot show that it is wrong. He has put forward an alternative theory which, also, does not carry widespread conviction. We are here on the borderland of scientific theorizing, and it is interesting to note that it is a region where æsthetic and philosophical predilections play an active part.

But whichever of these alternatives may be selected the world of physics still remains mysterious. For there is a whole vast region of phenomena which still remains outside these grandiose unifications. These geometrical schemes deal with matter in bulk. It is when we come to the ultimate constituents of matter that we encounter laws of an entirely different kind. And these laws, it would appear, are the really fundamental laws of the physical universe. They are of so singular a character that they cannot at present be made to form an integral part with the rest of our scientific knowledge. And they have given rise to speculations even more curious than those of relativity theory itself.

§ 5

The first step in the new direction was taken by Professor Max Planck in the year 1900. He was concerned with the manner in which heat is radiated by a hot body. Radiant heat, as we know, exists in the form of waves. They are just like light waves, except that they are too long to affect our sense of sight. But the heat radiated from a hot body is not all of the same wave-length. It is made up of groups of waves, some long, some medium, some short. Planck was interested in the question of which waves carried the most energy. Does most of the heat run into the short waves, or the long waves, or the medium waves? This question could be answered, of course, by experiment. It could also be answered by mathematical calculation. But it was on comparing these two answers that Planck's interest in the problem was excited. For they were in the most complete and flagrant contradiction with one another. This contradiction was no light matter, for the mathematical calculations rested on laws that had been universally accepted, and that were daily verified in countless experiments. Also, although technical errors were looked for in the calculations, none could be found. The heat measurements were repeated, and found to be correct. The contradiction was complete.

Planck found a solution of this puzzle by supposing that energy is not emitted continuously, but in jerks, or *quanta*. An atom radiating energy would radiate it in separate little lumps, as it were. Each of these little lumps can be regarded as an *atom of energy*. When Planck developed this conception mathematically he was able to reach very close agreement with the experimental results. In spite of this, however, the theory was felt to be so strange that it was not regarded favourably by the scientific world. And yet the theory, as put forward originally by Max Planck, was by no means so revolutionary as it presently became. For Planck did not assert that energy itself had an atomic constitution. He thought there was some sort of mechanism in the atom which caused it to emit energy in little separate lumps.

The further idea, that energy actually is atomic, came from Einstein. This idea is truly extraordinary. Energy most certainly seems to us to be one of those things that are necessarily *continuous*. In this it resembles speed, distance, time, and many other concepts. We can, in imagination, make any of these things increase as gradually as we please. Yet there are hints that the notion of atomicity may have to be applied to all these things. It may be that time, for instance, can only progress by little finite jumps. It may be that we shall have to conceive of space as composed of points not less than a certain finite distance apart. Although it may be mathematically possible to formulate these ideas, they are very sharply opposed to our intuitions. We have seen, however, that modern scientific men are not disposed to pay very much attention to our intuitions. The atomicity of energy is a concept quite in keeping with the modern tendency. An electric charge used to be considered as the distribution of a continuous fluid over the surface of the charged body. We now know that it is a collection of little separate charges—it is a collection of atoms of electricity. Matter, electricity, energy, they have all been atomized. This tendency is likely to spread.

As we have said, the scientific world was not at first disposed to regard Planck's extraordinary idea favourably. But a great variety of phenomena were discovered that lent support to it. One of the strangest and most puzzling of these is called the photo-electric effect. It is found that when light is allowed to fall on any substance electrons are emitted from that substance. The velocity of the electrons that come off can be measured. It is here that the phenomenon becomes puzzling. For the velocity of the electrons emitted does not depend at all on the intensity of the light. A powerful and highly concentrated beam produces more electrons, but they do not move any faster. If, however, we change the light, using light of a shorter wave-length—say, blue light instead of red light—then, however feeble the beam of light we use, the electrons come off with greater velocity. The shorter the wave-length of the incident light the higher the velocity of the emitted electrons. It does not matter how feeble the light is.

To see how utterly opposed this fact is to our ordinary notions we will illustrate it by an analogy due to Sir William Bragg. X-rays, since their wave-lengths are much less than those of ordinary light, cause electrons to be emitted at immense speed. Now X-rays are produced by the sudden stoppage of electrons as when, in a vacuum tube, we fire off a sheaf of electrons which are stopped by the walls of the tube. Each electron, thus suddenly stopped, generates X-rays, which spread out in an expanding sphere round the point of impact. As the spherical wave spreads out it naturally gets feebler and feebler, since the same total energy is spread over a wider and wider front. Yet, however feeble it may become, a portion of this wave, on striking something, say a metal, projects electrons from it having the velocity of the original electron that produced the X-rays. As Sir William Bragg says:

> It is as if one dropped a plank into the sea from a height of 100 feet, and found that the spreading ripple was able, after travelling 1000 miles and becoming infinitesimal in comparison with the original amount, to act upon a wooden ship in such a way that a plank of that ship flew out of its place to a height of 100 feet.

As Sir William Bragg's analogy makes clear, these facts are altogether opposed to our idea that light energy is uniformly distributed over a wave-front. But if we suppose that each stopped electron in the vacuum tube shoots out a little bullet of energy that arrives intact at the metal, then the results lose their paradoxical character. It appears that we must abandon the wave theory of light and adopt a bullet or corpuscular theory. But the distressing fact is that we cannot abandon the wave theory of light. Some of the very best attested experiments in science simply require the wave theory for their explanation, and cannot be explained by the corpuscular theory. But, as we shall see later, scientific men have evolved a curious sort of combined theory. And this theory is concerned not only with the propagation of energy, but also with the constitution of matter.

It was found that the theory of the atom as a miniature solar system that we have outlined also required the Quantum Theory for its explanation. In its original form the theory was impossible. According to the accepted laws of nature the revolving electrons in the atom should approach ever closer to the nucleus until finally the whole atom must vanish in a flash of radiation. So that on this theory of matter, matter could not exist. Nevertheless, the experimentalists were convinced that something corresponding to this picture must be true. Their experiments confirmed it. Again we had a contradiction between experiment and calculation. And again the contradiction was resolved in the same way, that is, by changing the basis of the calculations.

This was done by the brilliant young Danish physicist Niels Bohr, who was the first to apply the quantum theory to the atom. In its application to the atom the idea that energy can only increase or decrease by finite jumps resulted in saying that the revolving electrons in an atom can only move in certain orbits. These orbits are spaced at certain definite distances from the nucleus. There is a first orbit, second orbit, third orbit, and so on. An electron can revolve in any one of these, but it cannot possibly revolve in the space between two orbits. In particular it cannot get nearer to the nucleus than orbit No. 1. These steady orbits, we may mention, are absolutely impossible on the old mathematical theory. According to the old theory a revolving electron is continually radiating energy, and, so far from the spaces between orbits being sacrosanct or inviolable, the electron would infallibly pass through them on its journey down to the nucleus. According to Bohr, however, an electron only radiates energy when it jumps from an outer orbit to an inner orbit (which it does, in some unimaginable way, without passing over the intervening space). While it revolves steadily in one orbit it radiates no energy.

This theory, it must be remembered, is, in a sense, a purely arbitrary theory. No reason whatever can be given why the atom should be constituted like this. It can be shown that this construction will fulfil the requirements of the quantum theory, and that the older theory will not even account for the fact that matter exists. But the theory had no real rational basis; its justification was that it worked.

For a time it worked brilliantly. In particular it was very successful in explaining, in detail, the light radiated from glowing hydrogen. Every substance sends out its own distinctive kind of light, usually exceedingly complex, and one of the chief duties of any theory of the atom is to explain these complexities in detail. Besides explaining the light from hydrogen, Bohr's theory gave a good general account of the light emitted by other elements, but the mathematical difficulties presented by these more complex atoms did not admit of such detailed treatment. Nevertheless, the theory was considered to be satisfactory as far as it had been carried. But since then new experimental results have entirely changed the whole aspect of things. The Bohr theory has been definitely abandoned, and an entirely new light has been thrown upon the nature of matter.

Matter has revealed itself as being so extraordinary, even incomprehensible, a thing, that the resultant change in our outlook has been described as the greatest revolution in the history of scientific thought.

To put it briefly, the electron, the ultimate constituent of matter, is found to have the properties both of a wave and of a particle. The old particle conception, as used by Bohr, is altogether inadequate. And with this discovery the last possibility of picturing the electron has gone. It is strange that the more we investigate the true nature of matter, the most familiar thing in our experience, the more elusive it becomes.

The first man to initiate the new wave theory of matter was Prince Louis de Broglie, and it has since been developed by Schrödinger and others. But we shall postpone the consideration of their investigations until we have described the experimental evidence for the two-fold character of matter.

That electrons are particles is plainly shown by causing them to bombard a specially prepared screen. If we take a sheet of glass and lightly powder it with zinc sulphide crystals and then bombard it with electrons, we shall find that scintillations appear irregularly all over it. Each electron, on striking the screen, causes a faint spark which can be seen in the dark by the help of a magnifying lens. The stream of electrons behaves like a shower of rain falling on the screen, each impact giving a tiny spark. Here we have a clear and convincing proof that electrons travel like little bullets.

Now let us consider another experiment. When X-rays were discovered there was some doubt as to their nature. It was thought that they might be little particles; it was also thought that they might be waves. The matter was settled by causing X-rays to pass through crystals, and examining the patterns produced on a photographic plate. If that pattern consists of alternate dark and bright bands, then X-rays must be waves. For the dark bands are produced, it can be shown, by the crest of one wave coinciding with the trough of another, thus cancelling one another, while the bright bands are due to the fact that the crests of two waves coincide. Nothing but a wave motion can explain this phenomenon, and when the X-rays were found to give this series of bands it was unhesitatingly concluded that they are waves. It was on this kind of evidence that the wave theory of light was founded. What would happen if this experiment were performed with electrons? Judging from the scintillating screen experiment we should merely get an irregular distribution of electrons over the photographic plate. But when the experiment is performed we get a perfectly regular sequence of bright and dark bands, as in the X-ray experiment. Electrons, then, behave like waves.

This experiment is one of considerable delicacy. X-rays have considerable penetrating power, but the penetrating power of electrons is very slight. The electrons were fired through a thin sheet of metal, for an ordinary metal consists of microscopic crystals held together by a sort of cement. But owing to the feeble penetrating power of the electrons, the metal sheet had to be extremely thin. Professor G. P. Thomson prepared sheets of metal only one-millionth of an inch in thickness. Such metal sheets are very transparent. Objects can be seen through them quite easily. Even the bricks of a distant building can be counted if the sheet is held close to the eye.

The bright and dark bands obtained in this way are in the form of concentric rings, as in the X-ray experiments, but they are rather smaller than the X-ray rings. This indicates that the electron waves are shorter than X-ray waves, which is in entire agreement with de Broglie's theory.

We see, then, that the experimental evidence shows that electrons are both particles and waves. The notion of a particle is straightforward enough, but the waves that constitute an electron are not straightforward waves. In the system of waves constituting an electron there is a "disturbed area" which moves more slowly than the train of waves. We see something similar in a storm at sea, where there are patches of intense disturbance moving comparatively slowly. The advancing waves overtake and move through these patches. These disturbed areas reveal the position of the electron. But we cannot say that the disturbed area *is* the electron. For any such area has a tendency to spread, and if the matter of our world consisted of a number of disturbed areas it would by now have spread indefinitely. The disturbed area is not the electron, but the electron is somehow intimately associated with it. The electron travels with it, for example.

But when we come to give some sort of physical body to this theory we find it quite impossible to do so. We cannot imagine these waves to be physical disturbances in some sort of universal medium—an ether. They are, it appears, completely immaterial waves. They are as immaterial as the waves of depression, loyalty, suicide, and so on, that sweep over a country. Such immaterial waves are not necessarily mere metaphors. A wave of emotion sweeping over a country, rising and falling, passing from place to place, could conceivably, if it were measurable, be expressed as a wave. Similarly, these electron waves are sometimes described as waves of probability. If the waves are intense in any region, it means that the electron is likely to be there. But that these waves, whatever they may be, are completely unsubstantial, is shown when we come to apply this theory to atoms.

As long as we are considering a single electron we may be misled into supposing that its waves are physical realities, for they occupy three dimensions, and therefore our three-dimensional space could accommodate them. But we find that the waves associated with two electrons require a six-dimensional space to exist in, those associated with three electrons require nine dimensions, and so on. Each electron, in fact, requires a three-dimensional space to itself. This would seem to make it obvious that these waves are merely a mathematical device. It is distinctly disconcerting, therefore, to find that experiment, as we have seen, confirms their existence. The apparent agreement between calculation and experiment seems to be in some sense illusory. It is very difficult to avoid the conclusion that the experiments have not yet received their right interpretation. And there, for the present, we must leave the matter.

Enough has been said to show how different the scientific world has become from that plain, straightforward, objective universe with which we started. At the present day the scientific universe is more mysterious than it has ever been before in the history of thought. Although our knowledge of natural processes is greater than it has ever been, this knowledge is, in a way, less satisfactory, for in every direction we are faced by ambiguities and contradictions. Yet this state of affairs is not depressing; it is, on the contrary, extraordinarily stimulating. For we feel that our difficulties can only be resolved by man's imagination rising to the task of constructing radically new concepts. We have reached one of the great stages in the adventure of thought. We are, very possibly, already on the brink of some great new abstraction. We are required to see the universe with new eyes, and it is because it makes such demands and also holds out the promise of realizing them, that the study of science is so supremely worth while.

Perhaps the most important new concept that has yet emerged out of the rich and fascinating quantum theory is the Principle of Indeterminacy. This principle is the negation of the strict determinism that has hitherto reigned in science. Until quite recently science has assumed that a knowledge of the present is sufficient to enable us to determine the future. From a knowledge of the present positions and velocities of the heavenly bodies, for instance, we can prophesy future eclipses. There are many cases, of course, where we cannot say exactly what will happen, but that is because our knowledge of the conditions involved are not sufficiently precise. Thus we say that the result of tossing a penny or throwing dice is a matter of chance. But nobody doubts that the result is, in fact, exactly determined by the velocities, the resistance, and so on, involved. A full knowledge of all these factors would enable us to prophesy the result correctly. Again, in a specimen of radium only a certain percentage of radium atoms are breaking up at any one time. We could not say, of any particular radium atom, exactly when it will break up. It might break up within the next minute, or within the next thousand years. But the reason for our uncertainty is, we suppose, our insufficient knowledge of the constitution of the radium atom. We believe, or, rather, we have hitherto believed, that every event in nature is an example of strict cause and effect. The possible exceptions are provided by those of our own actions which are, we say, the result of "free-will." But many have thought that the strict determinacy found in nature would ultimately be extended to our mental processes, and that our consciousness of free-will is an illusion. The principle of indeterminacy has profoundly changed this outlook.

The principle of indeterminacy is founded on the fact that we cannot observe the course of nature without disturbing it. This is a direct consequence of the quantum theory. The effects are most obvious in the case of electrons, so we will confine our attention to electrons, which are, it will be remembered, the fundamental constituents of the material universe.

An electron in complete isolation would be unknowable. It is only when an electron is interchanging energy with some other part of the universe that we can become aware of its existence. But no interchange of energy can take place that does not involve at least one quantum or atom of energy. Now the electron is so small and light a body that this amount of energy is sufficient seriously to disturb it. Let us suppose, for example, that we are observing the electron's position and velocity through some sort of super-microscope, so as to calculate where it will be a second hence. In order to see the electron we shall have to use light. At least one quantum of energy will be involved, and this is sufficient to disturb the electron in an unpredictable manner. By using light of long wave-length, and therefore of little energy, we would not greatly disturb the velocity of the electron, but its position would be very hazy, since we cannot distinguish objects that are small compared with the wave-length of the light we are using. If, on the other hand, in order to determine the position accurately, we used light of very short wave-length, its great energy would hopelessly disturb the motion of the electron. Now for accurate prediction we must know both factors; we must know both position and velocity. But these two factors are so connected that the more accurately we know the one the less accurately we know the other. And this uncertainty, the quantum theory tells us, can never be less than a certain amount.

The difficulty can be made even clearer if we consider the electron from the point of view of the wave theory. According to this theory an electron corresponds in some way to a "disturbed area," or wave-packet, as it is sometimes called. The correspondence is such that the position and velocity of the wave-packet will give us the position and velocity of the electron. We may regard the electron as being situated somewhere in the wave-packet. Now wave-packets may be large or small. Also, wave-packets continually spread, but small packets spread faster than large ones. Let us consider an electron that is associated with a large wave-packet, and let us ask what we know about its position and velocity. We see at once that the position of the electron is vague, for it may be anywhere in the wave-packet. Its velocity, also, is somewhat indefinite, for if it is in the forward part of the spreading packet it is moving faster than if it is in the hinder part which is spreading backwards. The wave-packet is moving as a whole, but the forward spreading part is obviously moving faster than the average, while the backward spreading part is moving more slowly than the average. But since large wave-packets spread slowly the uncertainty about the electron's velocity will not be very great. It will be less the larger the wave-packet. If, on the other hand, we are dealing with a small wave-packet, we see that the position of the electron will be fairly well defined, but, owing to the rapidity of spread of a small wave-packet, the velocity will be much more uncertain. This example shows once more that we pay for precision in position by uncertainty in velocity, and vice versa.

We see, then, that we cannot accurately determine an electron's future behaviour. Any observations we make in order to get the necessary data disturb the electron in an unpredictable way, and therefore our data become useless in the very act of obtaining them. What, then, becomes of the determinism hitherto assumed by physical science? It becomes a useless principle. We may say, if we like, that the motion of an electron *is* strictly determined, although we can never determine it. But of what use is that statement to a scientific man? Since, by hypothesis, he can never verify the statement, and since phenomena happen just as if it were not true, the statement is, to the scientific man, mere useless lumber. A determinist may say, as some of them do, that indeterminacy is unthinkable.

Strict cause and effect, they may say, is a necessity of thought. This is the old problem of free-will and determinism, and scientific men will probably be careful not to enter on it. But if we believe in the reality of free-will, we can accept indeterminacy as a fundamental principle. When Eddington says that something analogous to free-will must be put at the basis of physical phenomena, he means that the principle of indeterminacy is something more than a pragmatic rule, something more than a temporary mathematical device, and that it belongs to as fundamental a category of thought as does the notion of strict causality.

Both alternatives are equally possible. The question is, which of these principles does nature obey? And the answer we have obtained so far is that the ultimate processes of nature are not strictly determined. This theory has no difficulty in explaining the fact that in practice, when we deal with appreciable lumps of matter, nature exhibits strict cause and effect. For this apparent uniformity of nature is merely a statistical effect. The idiosyncrasies of the individual electrons and atoms in any perceptible piece of matter cancel out, as it were. Indeed, one of the real tasks of science at present is to deduce the laws that govern matter in bulk from the laws that govern its ultimate constituents. The reduction cannot be effected the other way round. It is the electron that is the key to the universe.

CHAPTER IV

The Scientific Account of Origins

ONE of the most interesting questions that can be put to science is the question as to the origin of the present state of things. We are born into a world of inconceivable diversity. The mere varieties of material substance themselves defy enumeration. And even more unimaginable is the variety of living things, plants and animals. We ourselves, as has long been remarked, are fearfully and wonderfully made. The immediate scene of this panorama is the surface of a sphere of land and water, rotating on its axis, and circling round an immensely huger sphere of blazing gas. Attending it are other spheres which, judged by our standards, are completely meaningless, since they support no life, and have no bearings on ours. And the whole of this system, our sun and its planets, is, it appears, infinitely insignificant in the universe of matter. In every direction in space blaze countless millions of other suns, apparently without any reference whatever to our existence. Can we believe that this state of affairs always existed? If not, how did it come about?

It is probable that man has to reach a certain level

of ease and culture before he can ask even this question seriously. For we cannot believe that the primitive cosmogonies, which asserted such things as that God took the sun, moon, and stars out of a box and hung them up, were intended as serious answers to a serious question. Some primitive peoples, however, took the questions a little more seriously, but not seriously enough to warrant any detailed examination of their theories.

The Greeks, however, elaborated a point of view which, with suitable modifications of detail, would still be held as a representative opinion. Plato says:

> The philosophers say that fire and water and earth and air all exist by nature and chance and none of them by art, and that the bodies which come next in order—the earth, sun, moon, and stars—have been created by means of these absolutely inanimate existences. The various elements are moved by chance and also by inherent forces according to certain affinities amongst them—of hot with cold, dry with moist, soft with hard, and according to all the other accidental mixtures of opposites which have of necessity happened. After this fashion has been created the whole of heaven and all that is therein, as well as all animals and plants and all the seasons. These come from these elements, not by any action of mind or of any God, or from art, but by nature and chance only.

This theory, that everything has come about by chance through the random action of "laws of nature," is probably still the most widely held theory amongst educated people. For many centuries, in Europe, this theory was opposed by the doctrine of special creation as taught in Genesis. Through the labours of modern scholars, however, the real meanings of the statements in Genesis have become so elusive that it would be difficult now to say precisely what the doctrine of special creation, as taught there, really is.

The first really scientific attempt to answer the question of the origin of things was published in 1796 by the French mathematician, Laplace. This was the famous Nebular Hypothesis. Its main idea had already been put forward by Kant but, owing to insufficient mathematical knowledge, his treatment of it was very faulty. Laplace assumed, to begin with, a huge mass of hot gas in a state of rotation. This assumption is not perfectly gratuitous, for the existence of such bodies is suggested by telescopic observation. As this mass of gas cooled it would contract and rotate faster. It would also assume a flattened form. There would come a stage in this process when the outermost ring of this flattened body would become detached from the main mass. This would happen when the centrifugal force of the outer rotating particles was just sufficient to balance the gravitational attraction drawing them towards the centre. The outermost ring, detached in this way, would, Laplace thought, condense and form a planet. The main mass would continue to contract and to increase its rate of rotation until a second ring was left behind, and then a third ring, and so on. Each of these would condense to form a planet. The remaining central mass is the sun. The planets, by rotating on their axes, would also throw off rings which, condensing in their turn, form the satellites.

Although this theory enjoyed an immense popularity with scientific men for many years, modern mathematical research shows, pretty conclusively, that it cannot possibly be true. It appears that the scale is altogether too small. A rotating mass of gas, only large enough to supply the matter of the solar system, could not throw off planets. The rings that Laplace imagines being thrown off would not, as a matter of fact, condense. The mutual gravitational attraction of their particles would not be strong enough to prevail against the tendency of these particles to fly apart. Such rings of gas would merely dissipate themselves through space. They could not condense unless the primitive nebula was millions of times bigger than the nebula imagined by Laplace.

There is evidence for the existence of such nebulæ. They are the "spiral nebulæ," of which many millions are known to exist. Photographs of these bodies show them in all stages of development, and we see that they are passing through a process very like that imagined by Laplace. The process is not exactly the same for, instead of throwing off circular rings, these great nebulæ throw out spiral arms. Along the spiral arms there occur points of condensation. But each of these points results, not in a body the size of a planet, but in a body the size of a star. Laplace's nebular hypothesis, if applied on a scale sufficiently large to make it mathematically possible, results, not in the birth of planets, but in the birth of stars. As a theory of the birth of stellar universes it is pretty generally accepted. The change in the scale of the original theory is, we see, gigantic. A spiral nebula contains, on the average, enough material to make several thousand million stars. For such bodies Laplace's theory is applicable. It does nothing to explain the origin of the solar system.

The theory which seems to be, on the whole, the best attested supposes that the solar system resulted from a very rare accident. It is supposed that the planets were torn out of the sun by the close approach of a passing star. If the star were sufficiently massive and its approach sufficiently close, huge tides would be raised on the sun, and ultimately a colossal cigar-shaped filament would be torn out of it. Centres of condensation would be set up in this filament, and each of these centres would separate off as a planet.

The evidence for this theory is quite good. In the first place the cigar-shaped filament, thickest in the middle, corresponds pretty well with the observed sizes of the planets. For, if we imagine the planets lying in a straight line from the sun, in the order of their distances, the largest planets, Jupiter and Saturn, occur about the middle of this line. In either direction from the middle, towards or away from the sun, the planets tail off in size—approximately, at any rate. They would fit fairly well into a cigar-shaped filament. At first the orbits of the planets would be very elliptical, not nearly circular, as they are now. And in the course of describing these elongated ellipses they would, at some period, pass very close to the sun. If they were still in a sufficiently plastic condition, the sun would act on them as the passing star had acted on the sun, and pull filaments out of them. These filaments would condense and so give rise to the satellites.

This theory may be pursued, quite convincingly, into details. Thus it is a curious fact that the two largest planets, Jupiter and Saturn, each have nine comparatively small satellites, while the earth and Neptune each have one comparatively large satellite. This may be explained by taking into account the different rates of cooling of these differently sized bodies. The mathematical theory shows that the more liquid a planet was at birth the less likely it would be to break up under the gravitational attraction of the sun. But, if it did break up, it would give birth to a comparatively large satellite. Now it is natural to suppose that the large planets, Jupiter and Saturn, remained gaseous longer than Neptune or the earth, and therefore each produced several small planets instead of one large one. Still smaller planets, Venus and Mercury, have no satellites at all.

An apparent difficulty arises with the other two planets. Both Mars and Uranus are something of a misfit in the cigar, being smaller than they should be. And they each have small satellites, Uranus having four, and Mars two. But this may be explained by supposing that they were the smallest planets to be born in the gaseous state, that their condensing power was therefore less, and that they therefore lost some of their material by dissipation. Thus the theory hangs together very well, without any undue straining of the interpretation.

The theory also explains the fact that the planetary orbits are now nearly circular. At the beginning, as we have seen, they were very far from circular. But we may suppose that, in the catastrophic break-up of the sun, a great deal of dusty, gaseous matter would be scattered throughout the surrounding space. The planets, in moving through this dust, would be moving through a resisting medium, and it can be shown that the effect of this would be to make their orbits gradually approach the circular form. Indeed, the present departure of the planetary orbits from the strictly circular form enables the time that has elapsed since the break-up of the sun to be approximately calculated. The calculation is very approximate, however, owing to our lack of knowledge of the distribution of the primitive dust, and only allows us to say that the break-up occurred some time between one thousand million and ten thousand million years ago. Much closer limits can be reached by other methods, as we shall see later.

§ 2

A probable figure for the age of the earth, in round numbers, is two thousand million years. For practically the whole of that time the earth has been mainly solid. It has been calculated that the earth must have solidified within fifteen thousand years of its birth. This is a very small fraction of its life. And ever since the solidification occurred, the temperature of the surface of the earth has remained practically the same as it is now. For it can be shown that heat from the interior could percolate only very slowly through the crust. By far the greater part of the heat came from the sun, and the sun's radiating power must have been very much the same then as it is now.

At first the solid rocks composing the crust would, of course, have been only a little below their boiling point, since they had only just cooled down enough to solidify. As they continued to cool they would contract violently, and enormous fractures would occur. Having reached their present temperature, the rocks, as we have seen, would not get any cooler. But the inside of the earth would go on cooling and contracting until finally the outer crust would be left suspended, like a huge arch, without any material to rest on. There would come a time when the crust could no longer stand the strain, and it would collapse. Having settled down in its new position, the process would begin again, and after a time the crust would again collapse. It is thought that six or seven of these great crumblings can be traced in the earth's history. In this way, it is claimed, the existence of mountains on the earth's surface can be accounted for. There are other theories of mountain formation, but none, perhaps, so satisfactory as this.

But although this theory can account for the existence of mountains, it cannot account so satisfactorily for some other features of the earth's surface. In particular, it does not account for the distribution of earth and water. The great difficulty is the Pacific Ocean, which covers half the earth's surface. Why should all the land be crowded into one hemisphere of the earth? One theory suggests that the Pacific Ocean covers the depression left in the earth when the moon was torn out of it. But this supposes that the earth had solidified at the time the moon was born, and mathematical calculation seems to show that the moon could not have been born from a solid earth. The disruptive force of the gaseous sun would not have been sufficient, however close to it the earth may have passed. If, on the other hand, we suppose the earth to have been liquid, then obviously the separation of the moon from it could not have caused a depression, since the fluid would at once fill it up.

Another theory suggests that the great depression of the Pacific Ocean may have been scooped out by the primitive tides. We must remember that the moon only gradually moved away to its present distance from the earth. Now the nearer the moon is to the earth the greater is its tide-raising capacity. It can be calculated that, in those early days, the moon was raising tides nearly three miles high. Supposing that there already existed some difference of level between the northern and the southern hemisphere, these tides would tend to increase that difference. For water flows with more rapidity in an already deep place, and so tends to deepen it further. But this theory, like some others that have been proposed, is not much more, at present, than a speculation.

The age of the earth, since it solidified, can be determined with fair accuracy by a number of different methods. The best method is a result of the discovery of the radio-active elements. Uranium is such an element. It gradually transforms itself, at a perfectly definite rate, into lead. The rate of this transformation, it is found, is not altered by any extremes of temperature or pressure to which we can subject it. We are justified in regarding this rate as invariable. Fragments of uranium are found in certain rocks, and have doubtless formed part of those rocks since they first solidified. Together with the uranium we find lead. Now we cannot at once conclude, of course, that all the lead we find in the presence of uranium has been produced by the disintegration of uranium. But it so happens that lead produced from uranium is somewhat lighter than ordinary lead. It is possible to say, therefore, of any piece of lead, whether or not it has been produced by uranium disintegration. From this we can calculate how long the uranium present in that rock has been disintegrating, and, since the uranium has been present in the rock ever since the rock solidified, we learn the age of the solid rock. Estimates of this kind tell us that the rocks cannot have solidified less than fourteen hundred million years ago. These estimates are based upon the contents of the oldest rocks known to us. It may be, therefore, that the earth is still older than this. Other considerations, however, make it unlikely that the earth could have been, say, twice or three times as old as this. A good round figure for its age, as we have already said, is two thousand million years. This figure also agrees very well with geological estimates.

§ 3

The subsequent history of our earth is the chief concern of geology. It is strange that the science of geology is of such recent growth. The existence of various kinds of material—chalk, limestone, clay, etc.—arranged in layers one on top of the other, would have suggested, one thinks, that the top layers were formed after the lower ones. But it is usually supposed that this reflection was left for William Smith to make, an English surveyor, who roamed about England at the end of the eighteenth century, constructing canals. It had, however, occurred to people before him, and James Hutton had already published his *Theory of the Earth,* in which he ascribes the different layers to a quiet orderly deposit over a long period, and not as due to one single catastrophic flood—the Deluge. But William Smith was the first to publish a coloured geological map and to discuss, in a satisfactory way, the distribution of fossils through the various geological layers.

The existence of fossils had long been known. We know now that fossils are the actual remains of animals or plants that have been buried by natural causes (even the cast of a fossil shell, that is, the impression it has left on the rock, is called a fossil), but for a long time they were supposed to be curious "sports" of nature. The fossil of a marine organism, for instance, found high up on a hill was, with the early notions of the creation of the world, a difficult thing to interpret. In the early part of the sixteenth century fossils were supposed by some Italians to have been formed in the hills by the action of the stars, a view which, prior to 1579, Leonardo da Vinci combated. Then the hypothesis arose of a plastic force or, according to Andrea Mattioli, a fatty matter capable of fashioning stones into organic forms. But the hypothesis which held its place longer than any other, and is not yet extinct among the unscientific, is that they were relics of the Mosaic deluge.

But William Smith showed that certain geological layers have each their characteristic series of fossils. Some of these fossils are like species now living. Others are perceptibly different. Smith showed that some members of a series are wont to occur also in the layer above, others in the layer below, others in all three. This persistence shows that changes in these particular animals and plants, at any rate, could not have been sudden. Smith also noticed that, the farther back we go, the more do the fossils differ from forms now living, until presently we come to layers where all the species are extinct, and differ markedly from anything living now. Going still farther back we meet still other extinct forms, so that by the time we reach the lowest layers many dozens of extinct animals and plants have been encountered, all merging gradually into those above and below. Deeper still we find the fossils becoming exceedingly rare, and finally we reach a depth where they seem to be entirely absent.

These facts, in themselves, very strongly suggest a progress in time, and the early geologists, Hutton and Lyell, showed that small physical changes, operating through long periods of time, could account for the formation of the different geological layers. In Dr. Watts's words:

> Their work showed that physical processes, similar to those which can be studied in operation in different parts of the world of the present, are sufficient to account for the nature of the rocks of the past and the phenomena presented by them; that rocks of all kinds are being worn down at slow but reasonable rates by rain, rivers, and the sea; that the debris is being transported and laid down layer by layer in lakes or on the sea-bed; that this sediment is being compacted into stony rock, lifted to

> form new land, deformed to produce the structures presented by the rocks themselves and by their aggregates in mountain systems, sculptured and given the relief of land surfaces, cut down or depressed beneath the sea to receive a new load of sediment; that continents, too, may have been built and mountain ranges erected without any violent catastrophe, comparable in any degree with the magnitude of the earth as a whole, but as the outcome of slow and continually renewed movements, each on a comparatively small scale; and that all this has proceeded without any disturbing effect on life such as was contemplated by the catastrophists, though it must necessarily have had its influence on the details of that life, and in providing or withholding facilities for its migration: that, in brief, the rocks which bear the record of the history of the earth are the product of an infinity of small causes operating through a vast range of time.

On the assumption that the geological layers have been gradually formed by processes such as are operating at the present day, it is possible to estimate, approximately, the time that has elapsed since the first rocks began to be formed. The maximum observed thicknesses of the various geological layers total 529,000 feet, a little over one hundred miles. The rates of deposition observed at the present day vary considerably according to the local conditions. Thus the rate for Great Britain may be taken as one foot in three thousand years, while for North America the rate is nearly three times as slow. On the other hand Egyptian deposits occur at the rapid rate of one foot every four or five hundred years. If we wish to get the same round figure of two thousand million years for the age of the earth that other methods have given us, we must assume that the geological layers have been deposited, on the average, at the rate of one foot per four thousand years. As we see by looking at the present figures for Great Britain and North America, this is not at all an unreasonable figure. Thus we may say that the age of the earth, obtained from purely geological considerations, is fully consonant with the estimates derived from entirely different sources and is, to that extent, a confirmation of the theories of the geologists.

We see that the geological evidence is that the earth's crust has, in the main, been formed in an orderly manner, and not by a succession of catastrophes. We have the same impression when we study the fossils contained in these layers. We have abundant evidence of a slow and orderly progression of animal and plant forms. The highest animals, for instance, the mammals, exist only in the newer rocks. Lower down we find reptiles and birds. Preceding these are amphibians, and preceding the amphibians are fishes. Beneath the fishes lies a vast thickness of rock containing only invertebrate animals. A similar progress is observed in the case of plants. There can be no question but that the more complex animals appeared on the earth later than the simpler ones, and it suggests itself that a process of *evolution* has been at work, whereby the simpler creatures have gradually given rise to the more complex. A detailed study of the geological records ought to show us all the steps of this process, and acquaint us with all the forms that living things have assumed.

But the difficulties attending the detailed examination are considerable. A large proportion of organisms will not be buried in conditions favourable for their preservation. Many of the bodies washed out to sea, for example, will be devoured by other creatures, and only their hard coverings would be likely to be preserved. Even when these are preserved, the influence of percolating water may dissolve them, or they may be destroyed by chemical and mechanical changes in the rocks themselves. Furthermore, layers which once formed part of the sea-bed may be raised above sea-level, and then the exposed portion is exposed to all the crumbling effects of weathering, so that it and its fossil contents are gradually dispersed. We must remember, also, that only a fraction of the earth's surface is exposed to observation in cliffs and quarries, etc., and that many of these exposed portions occur in countries not yet adequately investigated. It is possible, also, that some of the exposed portions have not been quarried in a way favourable to the exhibition of their fossil contents and, even where the contents are exposed, they often pass in the first instance through unskilled hands, and so are mangled and distorted by the time they reach those who could have interpreted their evidence.

We see, therefore, that it would be unreasonable to expect geology to provide us with a perfectly clear and detailed account of the development of living forms. Nevertheless, considering the necessary imperfections of the geological record, the amount of evidence it offers is quite surprising. This evidence makes it practically impossible to believe that all the various species of animals and plants have been separate creations. We are forced to conclude that species have been modified in the course of time, that they are not fixed, and that, on the whole, progress has been from the more simple to the more complex.

The general outlook that alone seems consistent with the facts is admirably put by Darwin when he says:

> If we admit that the geological record is imperfect in an extreme degree, then such facts as the record gives, support the theory of descent with modification. New species have come on the stage slowly and at successive intervals; and the amount of change, after equal intervals of time, is widely different in different groups. The extinction of species and of whole groups of species reappear when the chain of ordinary generation has once been broken. The gradual diffusion of dominant forms, with the slow modification of their descendants, causes the forms of life, after long intervals of time, to appear as if they had changed simultaneously throughout the world. The fact of the fossil remains of each formation being in some degree intermediate in character between the fossils in the formations above and

> below, is simply explained by their intermediate position in the chain of descent. The grand fact that all extinct organic beings belong to the same system with recent beings, falling either into the same or into intermediate groups, follows from the living and the extinct being the offspring of common parents.

We are forced to very much the same reflections when we study the *geographical* distribution of animals and plants, as Darwin goes on to say:

> Looking to geographical distribution, if we admit that there has been during the long course of ages much migration from one part of the world to another, owing to former climatical and geographical changes and to many occasional and unknown means of dispersal, then we can understand, on the theory of descent with modification, most of the great leading facts in Distribution. We can see why there should be so striking a parallelism in the distribution of organic beings throughout space, and in their geological succession throughout time; for in both cases the beings have been connected by the bond of ordinary generation, and the means of modification have been the same. We see the full meaning of the wonderful fact, which must have struck every traveller, namely, that on the same continent, under heat and cold, on mountain and lowland, on deserts and marshes, most of the inhabitants within each great class are plainly related; for they will generally be descendants of the same progenitors and early colonists. On this same principle, although two areas may present the

> same physical conditions of life, we need feel no surprise at their inhabitants being widely different, if they had been for a long period completely separated from each other; for, as the relation of organism to organism is the most important of all relations, and as the two areas will have received colonists from some third source or from each other, at various periods and in different proportions, the course of modification in the two areas will inevitably be different.

The evidence scattered both through time and space leads us to believe that species are not fixed. It is interesting, in this connexion, to note that the definition of a species has always been a difficult matter. Authorities often disagree as to whether two given creatures belong to the same or to different species.

Lamarck, who was the first to make the question of the constancy of species a matter of general scientific discussion, believed that there is a "natural sequence" for living organisms, and that, if we knew all the species that exist and that ever have existed, we should be able to form them into a sort of long ladder or chain so that, starting at the beginning, we would arrive, by a series of almost imperceptible steps, at the end. Lamarck held that all our schemes of classification of animals into distinct species are artificial, and this conviction led him to the idea that species are essentially fluid. He pointed out that domesticated animals differ very markedly from their wild originals, and that man, by selective breeding from a common ancestor, has produced extraordinary variations, as, for instance, the greyhound, the spaniel, the bull-dog. Everybody would say, judging merely from their appearance, that these animals belonged to different species, yet they have a common ancestor. Changes just as remarkable, said Lamarck, are constantly taking place in nature.

§ 4

Lamarck's theory of evolution, although at one time pretty generally discredited, has now been revived by a number of prominent biologists. According to Lamarck, changes in an animal occur through use and disuse. Organs which are specially exercised become specially developed. The need for this special exercise arises from the conditions in which the animal lives; thus a changing environment, by making different demands on an animal, changes the animal. The giraffe, for instance, has developed its long neck in periods of relative scarcity by endeavouring to browse on higher and higher branches of trees. On the other hand, organs that are never exercised tend to disappear altogether. The eyes of animals that have taken to living in the dark grow smaller and smaller, generation after generation, until the late descendants are born eyeless.

The great assumption made by this theory is that the effects of personal, individual effort are transmitted to the offspring of that individual. This is a doctrine that is very much in dispute among modern biologists. We shall refer to this controversy later. In the meantime we must describe the most famous of all biological theories, the theory which, whatever its present position may be, has, by the attention it has directed to the matter, caused the *fact* of evolution to be accepted by all biologists. That theory is Darwin's theory of Natural Selection.

We must first mention an apparently unrelated historical fact. At the end of the eighteenth century political theories were common subjects of discussion. The American Declaration of Independence and the manifestoes of the French Revolution made such topics as the "Rights of Man," "Natural Justice," etc. matters of general interest. Many philosophers taught that the day of complete liberty and complete equality for all men was about to dawn. The Rev. T.R. Malthus, mathematician and economist, reflected that this state of affairs would lead to a vast over-population of the world. In such a world, population would rapidly outgrow the means of subsistence and therefore, he argued, checks on population would be necessary if misery and vice were to be avoided. He expressed these ideas in a book called *Essay on Population*. This essay was read, years later, by two biologists, Darwin and Wallace, and it started the same train of thought in each of them.

Let us start from some simple arithmetical facts. "The common thrush," Professor MacBride tells us, "begins to produce eggs when it is one year old, and its average length of life is about ten years. Every year a pair of thrushes will rear two broods, each consisting of about four nestlings. Starting from the offspring of a single pair we find that if all survived and mated, at the end of the tenth year, that is, at the completion of the life cycle of the parents, they would have produced a population of 19½ million. These in another ten years would grow to nearly 200 million million, and at the end of thirty years to about 1200 million million million. There would not be room for more than the one hundred and fifty-thousandth part of such an army of thrushes on the entire surface of the earth even if all stood side by side touching each other." Since, in fact, thrushes take up a very small part of the earth's surface, although they have been breeding for much more than thirty years, it is obvious that only a very small proportion of those born can survive and reproduce their kind. Similar remarks apply to innumerable other species of animals. Herrings, for instance, if allowed to proliferate undisturbed, would speedily choke up all the seas of the world. Darwin and Wallace were, of course, familiar with these facts, and it occurred to them that here we have Malthus's over-population problem in an exaggerated form. Malthus advocated "checks" to keep the population down. It is obvious that nature must apply checks of the most stringent kind. Of what nature are these checks?

They are varied, but the main ones are obvious when we reflect that the organism, whether plant or animal, must obtain food and avoid enemies. The success of the organism in doing this will in some cases be due to luck. We know that some seed falls on good ground, and some on stony ground. A falling slate from a roof may knock out the brains of a Newton as well as those of a cinema star. But such accidents do nothing to *vary* a species. The most they could do would be to exterminate it. But those individuals of a species which possessed some mental or physical advantages over their fellows, who were better adapted to the common environment, would, in the ordinary struggle for existence, and apart from sheer bad luck, have a greater chance of surviving and reproducing their kind. And if these advantages were of a kind that are inheritable, the species would tend to change.

That, briefly, is the simple idea of Natural Selection. Innumerable instances can be given. Thus it has often been recorded that lions, in attacking a herd of zebra or antelopes, invariably choose the weaker quarry. In general, as we should expect, strength and fleetness are advantages in the struggle for existence. This is so evidently what we would expect that it is hardly necessary to give instances of it. And we can easily see that there are many other characteristics, a greater or lesser measure of which would give their possessors a better chance of surviving.

The theory of natural selection thus rests upon two main facts, (1) that there are individual differences between members of the same species, and (2) that there is a struggle for existence. Each of these facts is incontrovertible. Offspring and parents are never exactly alike, a fact which makes selective breeding possible. And it is astonishing the varieties that selective breeding can produce, if continued over a long time, from small and haphazard differences. The amazingly various pigeons, for example, the pouter, the tumbler, and so on, are all descended from the ordinary rock pigeon. These varieties have been deliberately brought about by the breeder, who selects the points for which he will breed. He carefully mates those pigeons which have a variation in the required direction. The theory of natural selection states that a similar process occurs in nature. The struggle for existence takes the place of the human breeder. Nature sets a premium upon certain varieties as compared with others. Variations are continually occurring in all directions, as it were, and at random. Nature weeds out those that are unfavourable, and thus, little by little, a particular variation is established. It is supposed that in this way, given the vast periods of geologic time, every living and extinct species of animal and plant has been produced from some primitive form.

In this theory the origin of variations is left an unsolved problem. The theory merely states that they occur; it gives no reason for their occurrence. Further, Darwin, at any rate, assumed that the variations are always small, of the kind that we observe in daily experience. He also assumed that many of these variations occur entirely at random. But he also supposed that some of them come about through the purposive striving of the animal. He accepted, at any rate as a subsidiary factor, Lamarck's theory that the effects of use and disuse of organs can produce modifications which are inheritable.

The modern theory, what is called Neo-Darwinism, has been purified of certain of these assumptions. The notion that the effects of use and disuse can be inherited was attacked by Weismann. He pointed out that the reproductive cell of an organism is derived solely from the reproductive cell of its parent. None of its characteristics depend upon the rest of the parent's body. The germ-plasm, as he called the substance of the reproductive cell, passes without breach of continuity from generation to generation. The various bodies which contain it, in its passage down the ages, are mere vehicles or sheaths for it. Nothing that happens to any particular body in this chain is transmitted to any subsequent body, unless the happening is of such a kind as to influence the germ-plasm. Weismann cut the tails off generation after generation of mice, but the subsequent generations of mice were all born with tails, and the tails showed no tendency to get shorter. It appears that the cutting-off of a tail does not influence the germ-plasm. Mutilations, at any rate, are not inheritable. That section of the human race that practises circumcision finds that the practice must be continued in every generation. There is no tendency for the effect to become inheritable.

More recent experiments have avoided the method of mutilation, and have studied the effect upon the organism of varying the environment in such ways as may be expected to occur in nature. Changes in the organism can be brought about in this way, but it is still a matter of controversy as to whether such changes are inheritable. The reader who is concerned to accept only the definitely established conclusions of science must regard the question of the inheritance of acquired characteristics as being still open.

It may be pointed out, incidentally, that Neo-Darwinism, by denying that the purposive striving of the organism has any influence on its evolution, makes the whole of evolution a mindless process. That is, so far as the mind incorporated in living organisms is concerned. It is still possible to hold, of course, that a Mind, a Designer, is controlling the course of evolution, but the means employed seem to be purely mechanical. Inheritable variations occur quite independently of any purpose or striving on the part of the organism. It is in this sense that variations are said to occur "at random." It may be, nevertheless, that a detailed study of the way variations come about will reveal "purpose," although this purpose is not expressible in terms of the personal ambitions of the organism.

§ 5

The mechanism of heredity, about which a good deal is now known, was first illuminated by the researches of the Abbé Mendel. These researches were neglected for about half a century and were then encountered, almost simultaneously, by three independent biologists. To the student accustomed to the atmosphere of the "exact" sciences these researches, and the developments that have grown out of them, are quite unusually congenial. We here move in the clear and pleasant atmosphere of precise quantitative results and we have, underlying them, a theory at least as clear as the atomic theory in its early stages and one which, besides accounting for the known facts, leads to successful prediction. So much of biology, or at least of "natural history," presents us with an array of facts rather than gratifies our desire for comprehension. This is due, of course, to the sheer complexity of the subject and to the difficulty of isolating appropriate leading concepts. But in the matter of heredity, good working concepts have been found. They give no *ultimate* explanation, of course, any more than the atomic theory gave an ultimate explanation of the constitution of matter. And, in fact, the Mendelian theory is not unlike the atomic theory. The Mendelian theory asserts that inheritable characteristics are transmitted by discrete units. These units, from two parents, may exist side by side in the offspring, but they do not blend. We see at once that the laws of transmittance must be entirely different from what would occur in true blendings.

Let us consider, for example, a flower which exists in two varieties, red and white. The result of crossing these two varieties is to produce a pink flower. If these pink flowers are now bred between themselves, they do not produce, exclusively, pink flowers. They produce pink flowers, white flowers, and red flowers. We get the clue to the mechanism by which this occurs by studying the proportions in which these various colours are produced. In any large number we shall find that half the flowers are pink, a quarter red, and a quarter white.

The mechanism that would produce these results is very simple. Let us suppose, for example, that each flower, whatever its colour, contains two units of some kind which determines its colour. When two flowers are bred together, the offspring selects one unit from each parent. White flowers have two white-producing units. Red flowers have two red-producing units. When a white and a red flower are crossed, the offspring, which must take one unit from each parent, necessarily contains one red and one white unit. It is therefore a pink flower, and the crossing of a white and a red flower obviously can produce nothing but a pink flower. But now consider the result of breeding two pink flowers. Each flower contains a red and a white unit. The offspring may select a white from one and a white from the other. It will then be a white flower. Or it is just as likely to select a red from one and a red from the other. It will then be a red flower. Since each of these cases is equally likely we should expect to find as many white offspring as red offspring. But offspring containing one red and one white unit will occur just twice as often as either of the preceding cases. For the chance of selecting red-red is the same as the chance of selecting red-white. There will thus be a pink corresponding to every red. And the chance of selecting white-white is the same as the chance of selecting white-red. Thus there will be as many pinks as whites, for both red-white and white-red give pink. So we see that there will be twice as many pinks as there are either reds or whites. That is to say, one-half the progeny will be pink, and the other half will be split up equally into reds and whites. This agrees with the experimental evidence.

We see that these laws of inheritance are very different from those of pure blending. If, for instance, we cross a pink with a red we should expect, on the theory of pure blending, that the offspring would be a darker shade of pink. But on the present theory we see that each offspring must be like one or other of the parents, either an ordinary red or an ordinary pink. For the two units selected by the offspring must be either red-red or red-white. No other combination is possible. Similarly, the result of a cross between a pink and a white must be either a white or a pink.

The units we have discussed hitherto are supposed

to be of equal strength, as it were. A red unit and a white unit are equally matched, with the result that, when they are both present, the product is pink. But there are many observed results that cannot be explained by this simple hypothesis. We have to introduce units which are *dominant* and units which are *recessive*. An organism which contains a dominant unit and a recessive unit will not manifest intermediate characteristics. It will only manifest the characteristics belonging to the dominant unit. If, in our previous example, red had been dominant and white recessive, then the results of a cross would all have been red. Each of the offspring would contain one red and one white unit, but in each case the white would have been overpowered, as it were, by the red. But if we breed from this new generation of reds, we shall get both reds and writes as offspring. For since each parent contains a red and a white unit the offspring may select a white unit from each parent. The offspring will then, of course, be white. All the other offspring will be red, but some of them will be true reds, that is red-red, while the others will have the same units as our previous pinks, that is, red-white.

It was by studying the effects of units of this kind that Mendel was led to his theory. He experimented with two varieties of peas, Tall and Dwarf. "Tall" is a characteristic which is dominant to Dwarf. The result of the cross was to produce Tall peas, but these, when bred together, produced Dwarfs as well as Talls.

And of the Talls produced, some turned out to be true Talls and the others hybrid Talls.

We see that a recessive unit is a sort of lurking characteristic. Only when combined with another recessive unit can it manifest itself. It is stated, for instance, that, in the eyes of human beings, brown is dominant to blue. Thus a brown-eyed couple may give birth to a blue-eyed child. For it may be that neither parent is truly brown; each of them may be brown-blue. So that the child may select the blue unit from each parent. But it is evident that blue-eyed parents cannot give birth to a brown-eyed child. For blue eyes cannot exist unless they are blue-blue, unless they are blue all through, as it were. Thus there can be no brown units for the child to select. According to some authorities there are certain human abnormalities, some kinds of insanity and deaf-mutism, which are produced by recessive units. Thus it can happen that two normal people can produce an idiot child.

So far we have, in any particular case, dealt with only one characteristic, colour, height, and so on. But any organism has, of course, many characteristics. Each of these characteristics obeys Mendel's laws, and they are inherited independently of one another. Thus the possible combinations that can result from crossing, when we take into account several characteristics, can be very numerous.

The physical basis of these laws has been discovered. It has long been known that every living thing is built up out of "cells," and each cell contains a number of microscopic bodies called *chromosomes.* The cells of each species of organism contain their own distinctive set of chromosomes. A cell of the human body, for example, contains forty-eight chromosomes. For some creatures there are only two chromosomes, for others some hundreds. And chromosomes differ in size and shape.

Cells grow by division. A cell splits in half and becomes two cells. This process is accompanied by a number of very complicated internal phenomena, but the upshot of it, from our present point of view, is that each daughter cell emerges with the same chromosome outfit that was possessed by the original cell. This is secured by the fact that each chromosome, while the original cell is dividing, splits in half along its length, one half going to one daughter cell and the other half to the other. An exception occurs when the reproductive cells are being made. In that case the chromosomes of the original cell do not split in half, but half the total number goes to one daughter cell and half to the other. Thus the human ovum and the human spermatozoon each contains twenty-four chromosomes. When the ovum is fertilized, the spermatozoon adds to it its own stock of choromosomes, and thus the fertilized cell contains its proper outfit of forty-eight chromosomes, half being derived from the female and half from the male. This is the mechanism of heredity.

In speaking of the "mechanism" of heredity we are

using a term that is commonly employed by biologists. One must not conclude, however, that any mechanical explanation of these phenomena has been given. As a matter of fact, a mechanical explanation has not been even remotely approached, and some biologists believe that such an explanation is impossible. These biologists maintain that the sciences of life must use concepts peculiar to themselves, and that the concepts found adequate for physics and chemistry are inadequate for the phenomena of life. There has always, of course, been a school of biologists who hold that life cannot be explained on mechanical principles. Amongst them we must reckon the vitalists who asserted that a vital force, besides the forces known to physics and chemistry, plays a part in living phenomena. Vitalism is now discredited, however, but it does not follow that mechanics will be found sufficient. It is the less likely in view of the fact that mechanics has been found insufficient even in physics itself. A really adequate group of concepts, applicable to biological phenomena, has not yet been isolated.

Chromosomes are not simple bodies. Each chromosome is supposed to contain a chain of still smaller bodies, called *genes,* and these genes are the real units of heredity. Experiments, in particular the famous series of experiments carried out by Professor Morgan and his school on the fruit-fly *Drosophila melanogaster,* seem to make it clear that each gene is responsible for a definite characteristic. On the evidence afforded by a prolonged series of experiments Professor Morgan has constructed a map showing the distribution of the genes in the chromosomes of the fruit-fly, and against each gene is the particular characteristic for which it is responsible. Colour, shape of wing, presence or absence of eyes, and various other characteristics have been mapped in this way. In some cases a single gene is sufficient to produce a very marked effect. In other cases the effect is produced by a number of genes conspiring together. It can happen, also, that different genes counteract one another. It cannot yet be definitely asserted that all inheritance depends on genes. But the possible exceptions are comparatively few, and it seems clear that in nearly all cases the genes play at least the predominant role.

If we suppose that all inheritable characteristics are conveyed by the genes, we see that the theory of evolution requires us to suppose that changes in the genes can occur. The various theories of evolution, therefore, are really theories as to how these changes take place. There is no doubt, to begin with, that changes do occur. Sudden alterations, in both animals and plants, have often been observed. These alterations are not mere variations, such as are produced by the ordinary shuffling of the genes; they are much more radical and testify to an actual change in the germ-plasm itself. They are called *mutations.* A number of biologists, perhaps the majority, believe that mutations occur at random. Others believe that individual striving, on the part of the animal, can affect its germ-plasm, or that environmental influences can produce changes in it.

There is also the theory called Orthogenesis which states that the germ-plasm, in particular cases, at any rate, is predetermined to develop in certain ways. This development goes on quite irrespective of whether or not it is advantageous to the organism concerned. And there are the vitalistic theories, of which Bergson's *Élan Vital* is the most celebrated. We here have the course of evolution attributed to a purposeful striving manifesting itself in matter. This purposeful striving is not the God of the old theologians. It is not something directing the evolutionary process from outside, it is something embodied and manifested in the world of living things, and in ourselves it has reached consciousness. None of these theories has yet received general assent and, in this sense, the contribution of biology to modern thought is not yet unambiguous. So far as biology is concerned, Plato's summary of the Greek theory that everything came about by "chance" is still a possible, but not a necessary, belief.

The theory of genes and mutations make the process of inheritance much more intelligible than it was before. It is not yet generally agreed, however, that in mutations we have found the actual raw material with which evolution has worked. Observed mutations, say some authorities, are not sufficiently profound to explain the origin of species. And it has been asserted that, of the mutations that have actually been produced under experimental conditions, practically all are detrimental to the organism concerned. In spite of the immense number of observations that have been made, and of the detailed experimental work that has been done in recent years, there is not yet unanimity of opinion as to the way, or ways, in which evolution has come about.

But as to the fact of evolution there is universal assent. A study of the world of living things, present and past, makes such a conclusion inevitable. On a broad sweep, the progress from simple to more complicated forms is too obvious to be disputed. There are instances, it is true, where nature seems to go backwards. It has happened that evolution seems to reach its limit in some particular direction, and that the end-products of the process become extinct, while some still surviving ancestral simple form starts off evolving on a new line. Such eddies in the main current are not always explainable, but, in general, the emergence of the different species of animals and plants in time fits in with the evolutionary hypothesis.

§ 6

The beginning of the evolutionary process raises a question which is as yet unanswerable. What was the origin of life on this planet? Until fairly recent times there was a pretty general belief in the occurrence of "spontaneous generation." It was supposed that lowly forms of life developed spontaneously from, for example, putrefying meat. But careful experiments, notably those of Pasteur, showed that this conclusion was due to imperfect observation, and it became an accepted doctrine that life never arises except from life. So far as actual evidence goes, this is still the only possible conclusion. But since it is a conclusion that seems to lead back to some supernatural creative act, it is a conclusion that scientific men find very difficult of acceptance. It carries with it what are felt to be, in the present mental climate, undesirable philosophic implications, and it is opposed to the scientific desire for continuity. It introduces an unaccountable break in the chain of causation, and therefore cannot be admitted as part of science unless it is quite impossible to reject it. For that reason most scientific men prefer to believe that life arose, in some way not yet understood, from inorganic matter in accordance with the laws of physics and chemistry.

There is also the hypothesis, held by a few distinguished scientific men, that life is as old as matter and, in that sense, has had no origin. Lord Kelvin thought that the germs of life may have reached our planet from some other world. We are to picture these "germs" wandering about space from the beginning of time, and developing into various forms of living things whenever they strike a favourable soil. This means, of course, that the life of the wandering germs can remain dormant for an indefinite time. Arrhenius maintained that this was possible owing to the low temperature and lack of water-vapour in interstellar space. It has been found that certain bacteria, kept at a temperature of –200° C., are still alive at the end of six months, and the seeds of certain plants have been found to be still alive at the end of a century. Loeb stated that there is no reason why spores should lose more of their germinating power in ten thousand years than in six months. But, naturally, the indefinite persistence of life in the conditions of interstellar space is not a matter that can be tested experimentally. If it is to be believed, it must be as an article of faith.

The hypothesis that life has developed from inorganic matter is, at present, still an article of faith, although various chemists have put forward various hypothetical accounts as to how they think it might have been done. The fact remains, however, that none of these processes have been reproduced in a laboratory, and so their possibility lacks experimental proof. This aspect of the question was best summed up by T. H. Huxley sixty years ago. Modern knowledge has nothing essential to add:

> Looking back through the prodigious vista of the past, I find no record of the commencement of life, and therefore I am devoid of any means of forming a definite conclusion as to the conditions of its appearance. Belief, in the scientific sense of the word, is a

> serious matter, and needs strong foundations. To say, therefore, in the admitted absence of evidence, that I have any belief as to the mode in which existing forms of life have originated, would be using words in a wrong sense. But expectation is permissible where belief is not; and if it were given to me to look beyond the abyss of geologically recorded time to the still more remote period when the earth was passing through physical and chemical conditions, which it can no more see again than a man can recall his infancy, I should expect to be a witness of the evolution of living protoplasm from not living matter. I should expect to see it appear under forms of great simplicity, endowed, like existing fungi, with the power of determining the formation of new protoplasm from such matters as ammonium carbonates, oxalates and tartrates, alkaline and earthy phosphates, and water, without the aid of light. That is the expectation to which analogical reasoning leads me; but I beg you once more to recollect that I have no right to call my opinion anything but an act of philosophical faith.

It may be that some light will be thrown upon the general question by the further investigation of what are called the "filter-passing" organisms. These organisms are too small to be seen by the microscope, and they reveal their presence only indirectly. They are the cause of certain diseases. The question has been raised as to whether they are "alive" or not, and it seems that the question can be answered both ways. In fact, their characteristics reveal an unsuspected ambiguity of the term "life." It is possible that we have here a connecting link between what we ordinarily mean by living and not living matter.

In surveying this account of the way in which the world came to be as it is, we see that none of the chief hypotheses belong to the very highest class of scientific theory. None of them have the degree of validity of, for example, the electromagnetic theory of light. And it is unlikely that they ever will have this degree of validity, since they cannot be subjected to the test of experiment. The theory that the planets were born from the sun by the tidal effects of a passing star is such a theory. Even if instrumental power were so developed as to make such a phenomenon observable in some distant part of the heavens, the chances against its occurrence are so great as to make it very unlikely that it would ever be observed. Theories as to the early history of our planet are in a similar case; they deal with a set of conditions that cannot be reproduced. In this sense they may be called historical theories and, like history itself, they may reach a high degree of validity.

Theories of evolution do not altogether belong to this class. Evolution is presumably still going on, and is presumably due to the same causes that operated in past times. By selecting organisms which have a short life and which breed rapidly, such as *Drosophila,* and perhaps by speeding up the process by artificial restrictions, it may be that we shall actually observe the process of evolution. Observation of the geological records suffices to establish the fact that evolution has occurred; the definite determination of the method of its occurrence must presumably await the results of experimental investigation.

CHAPTER V

The Nature of Mind

THE order of the emergence of the different forms of life on this planet is made clear by the geological deposits. The last product of the process, to be found only in the most recently deposited rocks, is man—*Homo Sapiens.* Estimates of the antiquity of the various species of men vary, but we may say that the remains of modern man that have been discovered must be attributed to a period of from twenty thousand to fifty thousand years ago. The age of these fossil remains are determined from the geological strata in which they lie, the types of animal remains that may be associated with them, and the anatomical peculiarities that they present. In cases of ceremonial burial the implements buried with them also afford a valuable clue.

The best known of these races, belonging to our own species, is the Crô-Magnon, so called from the place where their remains were discovered in 1868. This race is completely modern. Indeed, it has been described as "in almost all respects the most perfect man physically that has come within our knowledge." The average height of the men seems to have been well over six feet, although the women were but little taller than those of the present day. Their brains, also, were larger than those of modern men, the brains even of the women exceeding the average of male brains of today. Besides various implements these men left behind them engravings, sculptures, and paintings, in and upon the walls of their caverns, which show a high degree of development. At about the same time there existed the Grimaldi race, much inferior physically, and with pronounced negroid features. There are various other remains scattered about Europe, perhaps belonging to still other races of Homo Sapiens, but their antiquity is disputed.

These remains, of men belonging to modern species, do not take us back, as we have said, more than fifty thousand years. But human remains of vastly greater antiquity have been discovered. They belong, as we should expect on the theory of evolution, to much more primitive types. Perhaps the earliest remains are those of *Pithecanthropos erectus,* discovered in the island of Java in 1891. A conservative estimate gives to these remains an antiquity of five hundred thousand years. In many physical characteristics this creature was distinctly human, but his brain, both in its size and in its characteristics, was very distinctly subhuman. Other remains have been found of still other species of submen, rather more recent than the Java man. The most famous of these is the Neanderthal man, short, powerful, not fully erect in posture, and altogether more brutish in appearance than modern man. The brain of this man was large, larger than some modern brains, but, unlike Crô-Magnon man, the higher parts of it were not well developed. Nevertheless, Neanderthal man was a skilled worker in flints, understood the use of fire, and buried his dead ceremonially. It is probable that he belonged to a species that died out completely, and that he is not in the line of descent that has produced modern man.

The remains of several species of primitive man have been found, but none of these are directly connected with ape-like forms. It is generally agreed, however, that the anthropoid apes and the various species of man had a common ancestor. The discovery of the various species of primitive man has done a great deal to reinforce this conclusion, although the actual line of man's descent from a definitely non-human stock is still a matter of speculation.

Human remains are rare, and from those alone it would be difficult to get any adequate idea of the distribution of man in prehistoric times. But a study of the tools and implements he left behind shows that, long before historic times, man had spread over the whole of the habitable world. The area he covered varied, sometimes expanding, sometimes contracting, for it so happens that man was contemporaneous with the great climatic fluctuations of the ice age, a period that began some hundreds of thousands of years ago, and lasted to within about twenty-five thousand years of our own day. The ice age comprises four cold epochs, when the ice spread from the polar regions right down to the temperate zone. Successive cold epochs were separated by periods when the climate was as warm, or warmer, than it is today. Man, as his remains proved, lived throughout all these alternations of climate although, naturally, he had to abandon regions as they became completely covered with ice.

This is the period that anthropologists call the Stone Age, from the implements that have been left behind. It is not suggested, of course, that the prehistoric men of that time did not also fashion tools from such substances as wood and bone, but the stone implements, being imperishable, are those that have survived. The earliest of these implements, consisting of roughly chipped flints, are so primitive that for a long time anthropologists were doubtful whether they were of human construction at all. Only very gradually did they assume a more artificial form. Indeed, many thousands of years elapsed before any very marked technical improvements were effected.

But at the time of the coming of Neanderthal man a distinct advance in implement making is perceptible. Their accomplishments, however, were primitive compared with those of Crô-Magnon man. This race made a variety of tools, stone, bone, and ivory, and ornamented them. Also, as we have seen, they began the fine arts, one of the outstanding phenomena of the whole prehistoric era.

Following on the unexplained eclipse of the Crô-Magnon race came a rather undistinguished period when no striking advances were effected, where, indeed, there was definite retrogression in certain respects, and then came the period of Neolithic culture when plants were cultivated and animals domesticated. Life became altogether more easy and agreeable. An interesting illustration of this fact is given by what we know of the diet of pre-Neolithic man. At one period, in Switzerland, the diet was ninety per cent cave bear, in Moravia ninety per cent mammoth, and in Denmark ninety per cent shell-fish. The Neolithic peoples, on the other hand, had fish, flesh, fowl, and garden products. Being cultivators, the Neolithic peoples naturally had fixed abodes. They built villages on piles, out over the waters of lakes or on their shores. The houses were built of wood, oak, beech, or birch, and contained separate rooms complete with fireplaces, etc. Looms also have been found, for Neolithic man understood the arts of knitting, spinning, weaving, embroidery, and the making of nets and baskets. He made boats, carved from tree-trunks, and it is probable that he made some use of the wheel. And we have evidence that he performed surgical operations, particularly trephining, with considerable skill and success.

It is evident that by this time man, the social animal, with an adequate language and a developing culture, had fairly come into being. The modern epoch of developing civilization, the distinctively human life, had begun. Indeed, recent researches suggest that the beginnings of civilization must be carried back a good deal earlier. There is evidence that the later types of primitive man were less primitive than has been supposed. But, whatever the correct time-scale of the process may be, there can be no doubt that modern man has developed from more primitive forms. This is true of his anatomy and seems to be equally true of his intelligence, while human societies have obviously developed enormously even within historic times.

At the time when Darwin stated his conviction that man had evolved physically from an ape-like ancestor many people were reluctant to suppose that this process applied also to man's mind. They thought that such a doctrine would have very undesirable repercussions on morality and religion. It took away from man his most distinctive and noble characteristic—his soul. So that some of them—among them Wallace, the co-inventor with Darwin of the theory of natural selection—while agreeing that man had been evolved physically, maintained that his mind and soul had been specially created. Nowadays this theory is felt to be very unsatisfactory.

The accepted doctrine, that there has been a gradual progression from an ape-like ancestor to modern man, raises the question as to when, during this process, the developing creature acquired a soul. If we suppose that the human mind and soul have not developed, but have been specially created, then any date we assign to this creation will appear arbitrary since the individuals so distinguished could have differed only almost imperceptibly, in all other respects, from their immediate predecessors. Also, actual psychological and anatomical observations made on the higher apes have done something to rob man of his unique position.

Nevertheless, the differences remain enormous. A great gulf separates even the lowest races of mankind from the highest living animals. We may suppose that the various fossil men exhibited different degrees of intelligence rising gradually from that of the anthropoid ape up to modern man. But such a progressive evolution of intelligence is, of course, conjectural. Within historic times there has been no development of human intelligence.

The general theory that intelligence has evolved fits in well enough, however, with actual observations. It is generally true that the physically simplest organisms have also the most rudimentary forms of intelligence. Indeed, if we define intelligent behaviour as successful adaptation to new conditions, then we must conclude that large groups of organisms are without intelligence. The moth which flutters towards a candle flame is responding mechanically to a stimulus; it is not displaying intelligence. These tropisms, as they are called, which cause an organism to behave in an invariable way without any regard to its personal advantage, cannot be classed as intelligent, although, in the creature's normal environment, they may be of advantage to it. The same may be said of reflex actions, as coughing, blinking, sneezing. Such reactions may be of advantage to their possessor, but they are not intelligent. It is characteristic of all these reactions that they are completely rigid and invariable. Instincts belong to a higher level of development, and although they can manifest a great degree of fixity they are, in the higher organisms, much more variable.

The question of the origin of this group of reactions is still a matter for speculation. A theory once held maintains that instincts are cases of "lapsed intelligence." It supposes that the original forms of life were able to choose, amongst the various forms of behaviour possible to them, those that were of advantage. These chosen forms of behaviour then became fixed habits and, as such, were transmitted to the descendants of the organism. They thus lost their original spontaneous character and became the automatic things we know.

This theory is at present regarded with disfavour. In the first place, it is very unlikely that the primitive intelligence it postulates ever existed. Further, the theory supposes that acquired characteristics are inherited—a supposition which is now, as we have seen, something in the nature of a heresy. The modern tendency is to account for all these reactions, tropisms, reflexes, instincts, by the old explanation of "random variations." Behaviour, it is assumed, is conditioned by physical structure, and random variations in structure give rise to corresponding differences of behaviour.

Favourable variations are preserved, in the usual manner, by natural selection.

The relation between physical structure and behaviour that is assumed by this theory has not been made out as clearly as could be desired. It is true that, on the whole, complexity of behaviour is associated with complexity of the nervous system. But the correspondence is not very uniform. Certain organisms, such as the ant, the bee, and the spider, display very complicated instincts, although their nervous organization is relatively simple. Other organisms, on the other hand, such as certain of the mammals, have extremely complicated nervous systems, but manifest much less remarkable instincts. Thus the correspondence between structure and inherited reactions is not yet very clearly made out.

If we define intelligence as the capacity to modify behaviour successfully in face of new situations, then it is a matter of common knowledge that some creatures are more intelligent than others. Certain theorists, it is true, have maintained that all animals are automata, and therefore have nothing that can properly be called intelligence at all. It is impossible to refute this theory by observation, since we have no direct access to the minds of animals. But then we have no direct access to any mind but our own. We attribute minds to other human beings on the strength of their behaviour, and we are justified in applying the same inference to animal behaviour.

Actual experiment shows that animals differ very considerably in their capacity to learn by experience. A common test is to give an animal two paths to choose from, one of which leads to food while the other does not. A creature of human intelligence, having once hit on the successful path, would stick to it. But such a creature as a frog, for example, seems almost incapable of profiting by its past successes. Very patient experimentalists have found, however, that even a frog can learn. Indeed, it cannot be definitely stated that *any* animal is altogether incapable of learning. On this ground alone, therefore, we should not be justified in saying that any animal's intelligence was radically different from our own. With the higher organisms, particularly the great apes, we have evidence of mental processes akin to the human. Such highly developed creatures as orangs and chimpanzees, faced by a new situation, have been known to make correct inferences for dealing with it, based on past experience, as by using a stick to draw in food otherwise out of reach, or by piling boxes on one another and mounting them in order to reach food suspended from the roof.

§ 2

Although the theory that mind has steadily evolved is perhaps the one that most naturally suggests itself, it is not necessitated by the observed facts. It is quite possible to suppose that sudden breaks have occurred, and that radically new elements have appeared. As we have seen, it has been maintained that consciousness is peculiar to man, and that no animals possess it. Even if we deny this, and insist on attributing consciousness to the higher animals, we might hesitate to extend it all down the scale and to attribute it, say, to the amœba or to plants. It might be difficult even to stop there, so that presently we should find ourselves regarding the whole inorganic world, also, as conscious. There is, perhaps, no conclusive reason why we should not do this, and there are philosophers who maintain that we should be correct in doing so, but it might be as well, before doing so, to search for alternatives. Any alternative will, of course, suffer from the disadvantage of postulating a breach of continuity.

This is the objection that is brought against Herbert Spencer's well-known attempt to describe the evolution of mind. He traces out the growing complexity of the nervous system, and shows how we may suppose the interplay of the nervous currents in the brain to have grown more and more complicated. And somehow, from the growth of complexity, consciousness arises. He says:

> These separate impressions are received by the sense—by different parts of the body. If they go no further than the places at which they are received, they are useless. Or if only some of them are brought into relation with one another, they are useless. That an effectual adjustment may be made, they must all

> be brought into relation with one another. But this implies some centre of communication common to them all, through which they severally pass; and, as they cannot pass through it simultaneously, they must pass through it in succession. So that as the external phenomena responded to became greater in number and more complicated in kind, the variety and rapidity of the changes to which this common centre of communication is subject must increase—there must result an unbroken series of these changes—there must arise a consciousness.

The chief interest of this passage is for the light it throws on Spencer's conception of an argument. That the conclusion is a complete *non sequitur* is apparent. If Spencer had intended it as a description of the actual circumstances in which consciousness arose, it would be no more mysterious than any other ultimate fact in nature. But as an *explanation* of the origin of consciousness it is, of course, entirely valueless. It is valueless because it involves an entire breach of continuity.

The old-fashioned materialistic philosophy, which explained a man's thoughts as consisting of the motions of little billiard balls in his head, suffers from the same defect. The transition from colliding spheres to thoughts is too abrupt. There is nothing in our knowledge of the properties of motion, or of little hard particles, which prepares us for the conclusion. Nevertheless, it may be that materialism, as a description of fact, is true. But unless our notions of hard particles and motion were greatly enriched, it would remain an irreducible fact.

The idea that materialism may be true as a description, although unintelligible as an explanation, fits in with the modern theory of Emergent Evolution, a theory to which certain philosophically minded biologists are according sympathetic attention. The theory, put briefly, states that at various stages of material complexity, radically new properties emerge. According to this theory both Life and Mind are emergent properties of certain material aggregates. A complete knowledge of the constituents of these aggregates would not enable us to predict that, in combination, they would manifest the properties of life or mind.

The general notion can be illustrated by considering a chemical compound. Water, for instance, is composed of hydrogen and oxygen. Can we predict, from a complete knowledge of the properties of oxygen and of the properties of hydrogen, the properties of water? We are here referring, of course, to the properties of hydrogen and of oxygen, taken in isolation. It is a fact, of course, that science cannot at present account for the properties of water from its knowledge of the properties of hydrogen and of oxygen, taken in isolation. But probably most scientific men would say that the reason for this is that our knowledge of the properties of hydrogen and oxygen is incomplete. The theory of emergence states that the reason lies deeper; the properties of water are not implicit in the properties of hydrogen and oxygen. A scientific archangel, brooding from eternity on hydrogen and oxygen, taken separately, could not say that their combination would manifest the properties of water. In the same way, the property of life which is manifested by certain material aggregates could not be inferred from any amount of knowledge of the constituents of those aggregates. And the same applies to the still higher complexities which manifest mental properties.

This theory is certainly not illogical, but it suffers from the disadvantage that it can never be proved. On the other hand, it could conceivably be disproved. As our knowledge of the hydrogen and oxygen atoms increases, it may be that we shall be able to deduce that the properties of water must be exhibited by their combination. The theory of emergence will then be disproved. But our failure, however often we fail, does not prove the theory of emergence—it only makes it possible; perhaps, as time goes on, probable. Thus our attitude towards the theory of emergence depends, to some extent, upon our estimate of the present level of scientific knowledge. Has the science of physics a great deal yet to learn about the properties of matter?

This is obviously a question that no man can answer, and scientific men, speaking officially, always say that science has barely begun. If this is true in the sense that the present principles and entities of physics are capable of indefinite application, then the theory of emergence must be regarded as, at best, a mere possibility, but not as a working hypothesis. The theory of emergence really assumes that our knowledge of physical and chemical entities is, in essentials, complete. Therefore, our failure to deduce the properties of a living cell from the properties of its constituents is to be taken as indicative of an actual breach of continuity in nature. Thus a completely unified science is, in the nature of things, impossible. We shall never reach a set of concepts in terms of which all phenomena can be described.

This statement is quite possibly true and, to many minds, will seem highly probable. But it is also possible that these difficulties in passing from the physical to the chemical, from the chemical to the living, and from the living to the mental, have an entirely different source. We must remember that the entities used in a science are abstractions from experience. And only a certain group of experiences are regarded as relevant. The entities with which a science works, and in terms of which it tries to account for the particular set of phenomena it is investigating, are all composed out of certain selected bits of our total experience. And they are composed as economically as possible. Scientific concepts are never any richer than they need be for the particular purposes for which they are designed.

Thus, for the purposes of the early kinetic theory of gases it was sufficient to regard atoms as little hard spheres. In the science of that day an atom *was* a little hard sphere. To the layman of that time science had "discovered" that matter is built up of little hard spheres, all perfectly homogeneous and of invariable size and weight. But as our experience of certain phenomena was extended and refined, it was found that this simple atom was not sufficient. It was "discovered" that it was more complicated than had been supposed. In course of time it became transformed beyond recognition. The modern atom, instead of being something with the fewest possible rudimentary properties, is a very richly endowed object indeed. But every step in this enrichment of the atom has been made as grudgingly as possible. Even now the atom is only as complicated as is necessary to explain the phenomena of spectra. The atom of the kinetic theory of gases would be hopeless for this purpose. Is it very surprising, then, that the atom of spectrum theory is not good enough to explain the phenomena of life? Presumably, if we want it to play that role, we have to "discover" more properties in it. By endowing the atom with a sufficient number of properties we shall make it adequate to explain, not only the phenomena of physics and chemistry, but also living and mental phenomena.

Nevertheless, it seems that the atom will then have undergone a real change. It does not seem that it can perform these new tasks without becoming something different in kind from what it was before. The atom of modern physics, different as it is from the Victorian atom, is different chiefly by being a much more complicated mechanism. It is true that our notions of mechanism have to be considerably extended to accommodate it, but such notions as mass, distance, motion, still remain fundamental in our construction of the atom. From this point of view we can say that the modern atom is not different *in kind* from the Victorian atom. But it seems impossible that any extension of this sort of complexity will help in explaining mental phenomena, even if it proves adequate to the phenomena of life. So that amongst the new properties with which we propose to dower the atom, we shall probably have to include a rudimentary form of consciousness. Perhaps there is a hint of this in the modern doctrine that the atom manifests "something like free-will."

The chief point in favour of the view we have just sketched is that it preserves the principle of continuity. Scientific men, as a whole, are very strongly in favour of this principle. It may be, however, that their feeling is no more than a prejudice. There are philosophers who believe that the world is a plurality, that it is composed of things essentially distinct. But the principle of continuity has led to so much fruitful work in science that it will probably long remain as a working hypothesis. It will be noted that this way of securing continuity, by postulating some rudimentary form of consciousness even for the ultimate particles of matter, involves a sort of dualism. For consciousness is something peculiarly different from the other fundamental properties attributed to matter.

We see that the question of the Evolution of Intelligence bristles with unsolved problems. We are not yet clear as to whether certain types of behaviour are to be classed as intelligent, nor can we draw any clear line of demarcation between the conscious and the non-conscious. The connexion between physical structure and mental characteristics is still very largely hypothetical. And we do not know whether consciousness arises only at a certain stage of complexity or whether it must be postulated of all living matter—or even of all matter. Thus the theory of evolution, although it can tell us a great deal about the development of our bodies, can tell us very little about the development of our minds. It may be that the structure of our minds is completely conditioned by the structure of our nervous systems, but researches on the nervous system, at present, throw practically no light on our mental processes. For the understanding of these we have to appeal to different methods.

§ 3

Unfortunately, the science which deals with mind, the science of psychology, is at present in a very rudimentary state. Indeed, it would be denied by many that any such science exists. Certainly there is no generally accepted body of psychological doctrine. There are, rather, a number of different theories, each having a limited range of application and, where they profess to deal with the same phenomena, differing profoundly from one another.

Amongst these theories the one with the greatest pretensions to being robustly scientific is Behaviourism. Behaviourism, in its strict form, says that what we call mental processes are in reality bodily movements. When we say that we are thinking of something, or perceiving something, this statement really means that our body is behaving in a certain way. It is at first very difficult to believe that these statements mean what they say. One finds oneself interpreting them to mean, for instance, that mental processes are always accompanied by bodily movements, or that bodily movements and mental processes are two indissoluble aspects of the same event. Without such interpretations the statements sound obviously absurd. But it appears that it is this obvious absurdity that constitutes the original element in Behaviourism.

Behaviourism gets rid of what we call "mind" altogether. We do not think; we make incipient speech movements. We do not perceive anything; we adjust our eyeballs. It would certainly seem to be a waste of time to discuss this theory were it not for the fact that there are a fair number of people who profess to believe it, as well as a large number of people who, for one emotional reason or another, would like to believe it.

The bodily movements assumed by Behaviourism are not necessarily overt movements. They include such things as slight movements in the throat, changes of blood pressure, and so on. A less pure form of Behaviourism, when evidence for these movements is not forthcoming, postulates unobservable molecular movements in the brain, and thus becomes a form of old-fashioned materialism. But even materialism has mostly acknowledged that there are mental processes accompanying, although wholly conditioned by, these molecular movements.

It is not logically impossible to deny that other people have minds, and in this fact consists such plausibility as Behaviourism possesses. We deduce the existence of other people's minds from their behaviour, and therefore we may say that their minds are, to us, hypothetical causes of their behaviour. The Behaviourist asserts that nothing is gained by postulating these hypothetical causes; we may confine ourselves wholly to the behaviour.

It is true that there is much in the behaviour of other people, as when we are listening to a conversation on politics, for example, which gives no evidence of thinking, and which suggests that purely automatic reactions are taking place. But it is more surprising to learn that the Behaviourist finds no evidence of thought processes within himself. The fact is, that each of us is immediately aware, within himself, of sensations, images, imagination, and reasoning. And we have not, in practice, the slightest doubt that we are correctly imputing these activities to other people. A compass-needle, in following the movements of a magnet, is certainly, according to the Behaviourist, "perceiving" the magnet. But we know that with our own perceptions, even if our bodily behaviour is as regular and automatic as that of the magnet, something more is involved. And this something more is, of course, the sensations, images, etc. of which we are directly aware. According to Behaviourism these sensations, images, etc. must all be reducible to bodily movements. They must be *identical* with bodily movements. That this is impossible has been explained very lucidly by Dr. Broad, who has throughout shown a truly admirable patience in his discussion of what he calls this "preposterously silly" theory:

> Let us suppose, for the sake of argument, that whenever it is true to say that I have a sensation of a red patch it is also true to say that a molecular movement of a certain specific kind is going on in a certain part of my brain. There is one sense in which it is plainly nonsensical to attempt to reduce the one to the other. There is a something which has the characteristic of being my awareness of a red patch. There is a something which has the characteristic of being a molecular movement. It should surely be obvious even to the most "advanced thinker" who ever worked in a psychological laboratory that, whether these "somethings" be the same or different, there are two different *characteristics*. The alterna-

tive is that the two phrases are just two names for a single characteristic, as are the two words "rich" and "wealthy," and it is surely obvious that they are not. If this be not evident at first sight, it is very easy to make it so by the following considerations. There are some questions which can be raised about the characteristic of being a molecular movement, which it is nonsensical to raise about the characteristic of being an awareness of a red patch; and conversely. About a molecular movement it is perfectly reasonable to raise the question: "Is it swift or slow, straight or circular, and so on?" About the awareness of a red patch it is nonsensical to ask whether it is a swift or a slow awareness, a straight or a circular awareness, and so on. Conversely, it is reasonable to ask about an awareness of a red patch whether it is a clear or a confused awareness; but it is nonsense to ask of a molecular movement whether it is a clear or a confused movement. Thus the attempt to argue that "being a sensation of so and so" and "being a bit of bodily behaviour of such and such a kind" are just two names for the same characteristic is evidently hopeless. And this is what the Behaviourist has really got to do.

Although Behaviourism, regarded as an ultimate account of mental processes, must be rejected, it has nevertheless been of some value. It has drawn attention to what we may call the mechanical aspect of some of the activities we call mental. It has gone as far as it can in reducing mental processes to reflexes, such as sneezing, blinking, etc., and, in so far as such reductions are successful, they give us information which is both new and important. Such reductions, however, do not usually lead to simple reflexes, but to conditioned reflexes.

The study of conditioned reflexes has come into great prominence of recent years, chiefly through the work of Pavlov. Pavlov's experiments have been conducted on dogs, but they deal with such basic phenomena that it is likely that they throw light on certain fundamental processes in higher animals, including human beings. At the sight and smell of food, saliva will flow into the mouth of a normal dog. If the dog has had its cerebral hemispheres removed, however, it will not salivate until the food is actually thrust into its mouth. In the second case we are dealing with a simple reflex action. In the first case the action is more complicated and involves a psychical response. The old-fashioned psychologist would call it an instance of the "association of ideas." The dog has learned by experience to associate that sort of smell and appearance with eating.

It occurred to Pavlov to try experimentally whether this one-link chain of association could be lengthened. Thus he would start by ringing a bell before presenting food to the dog. After a sufficient number of such experiences the dog would associate the ringing of the bell with the appearance of food, and would salivate when the bell rang and before the food had appeared. Such a reaction is called a "conditioned reflex." Pavlov rang numerous changes on this theme. He tried various stimuli, noises, colours, shapes, touches on various parts of the body, electric shocks, and so on, and found that the dog could learn to regard any such stimulus as the prelude to being fed. He also introduced stimuli which were *not* followed by food, and, when the dog had thoroughly learned to ignore these, he tried the effect of combining the two sorts of stimuli, the one that preluded food and the one that did not. In this way, besides advancing his main purpose, as we shall see, he obtained some highly interesting results concerning the sensory discriminations of dogs. They are so interesting that, although they are incidental, some examples are worth giving.

A dog would be taught to associate the appearance of a black patch with the appearance of food, and would be taught to ignore the appearance of a white patch. These reactions having been thoroughly established, the dog would then be presented with a grey patch. The dog easily distinguished this from a black patch, and did not salivate. Deeper and deeper shades of grey were presented to the dog in order to determine at what point his power of discrimination broke down. The interesting discovery was made that, in this respect, the dog's power of discrimination perceptibly surpassed that of human beings. Shades of grey so deep that human observers could not distinguish them from black evoked no salivation. In the discrimination of shapes, however, the dog showed himself inferior to human beings. Having learned to associate a circular shape with food, the dog was plunged into a frenzy of indecision on being presented with a broad elliptical shape which still differed perceptibly, to human eyes, from the circular form. In respect to sounds, however, the dog again showed a marked superiority to human powers of discrimination. He could detect very slight variations in loudness, he could hear sounds so high in pitch that they were inaudible to human ears, and he was extremely sensitive to rhythm, being readily able to distinguish the difference between a metronome beating ninety-six to the minute and one beating one hundred.

These facts about a dog's sensory discriminations are interesting, but their discovery was incidental to Pavlov's main purpose. We have seen that a dog can learn to ignore a stimulus. These inhibitory processes are, it appears, much more fragile and easily disturbed than are the excitatory responses. Inhibitory processes play a very large part in our psychological life. We should be quite unable to function as we do were it not that we are constantly inhibiting our responses to stimuli—even if the stimulus be no more than the ticking of a clock whilst we are reading. Pavlov found that, with his dogs, inhibitions have a tendency to spread. Thus a dog's positive reaction to a stimulus is somewhat weakened if he has just previously been subjected to a stimulus which he has been trained to ignore. But the details of the interactions between excitations and inhibitions are complicated. Thus it appears that, if one part of the brain is active, it tends to inhibit activity in other parts. Similarly, an inhibitory process in one part seems to cause increased activity elsewhere. Perhaps the most interesting generalization to which Pavlov has been led by his study of inhibition is his conclusion that sleep is a general form of inhibition. He says:

> Internal inhibition during the alert state is nothing but a scattered sleep, sleep of separate groups of cellular structures; and sleep itself is nothing but internal inhibition which is widely irradiated, extending over the whole mass of the hemispheres, and involving the lower centres of the brain as well.

These researches of Pavlov's doubtless leave a great deal of the psychology of dogs yet unexplained, and we should not expect, therefore, that they would throw very much light on the psychology of human beings. Nevertheless, the results obtained by this study are of particular interest because they have been obtained by a sound scientific method. In this respect they differ markedly from other and more ambitious accounts that have been given of our mental processes. This does not mean that the results achieved by other methods may not be more valuable, but only that we are inclined to have less confidence in them.

That conditioned reflexes can be established in children has been made clear by the experiments of Dr. Watson. A human baby is born with remarkably few unconditioned reflexes, but amongst these is fright at a loud noise. Advantage has been taken of this fact to obtain fright reactions to other things. Thus, if an infant is shown a teddy-bear at the same time that a loud noise is produced, the child will learn to manifest fright at the mere appearance of a teddy-bear. This response may become diffused, as it were, so that it takes place in the presence of almost any furry object. A conditioned reflex is established, exactly as in the case of Pavlov's dogs. It would be possible to illustrate this discovery by accounts taken from learned writings, but the most vivid account, and one which is scientifically impeccable, is given by a contemporary novelist [1] in his vision of the future world that may be created by science. In the new Pavlovian Conditioning Rooms bowls of roses and gaily coloured pictures of beasts, fishes, and birds, have been spread out. A batch of babies, ripe for conditioning, have been brought in.

> The babies at once fell silent, then began to crawl towards those clusters of sleek colours, those shapes so gay and brilliant on the white pages. As they approached, the sun came out of a momentary eclipse behind a cloud. The roses flamed up as though with a sudden passion from within; a new and profound significance seemed to suffuse the shining pages of the books. From the ranks of the crawling babies came little squeaks of excitement, gurgles, and twitterings of pleasure.

[1] *Brave New World,* Aldous Huxley.

The Director rubbed his hands. "Excellent!" he said. "It might almost have been done on purpose."

The swiftest crawlers were already at their goal. Small hands reached out uncertainly, touched, grasped, unpetalling the transfigured roses, crumpling the illuminated pages of the books. The Director waited until all were happily busy. Then, "Watch carefully," he said. And, lifting his hand, he gave the signal.

The Head Nurse, who was standing by a switchboard at the other end of the room, pressed down a little lever.

There was a violent explosion. Shriller and ever shriller, a siren shrieked. Alarm bells maddeningly sounded.

The children started, screamed; their faces were distorted with terror.

"And now," the Director shouted (for the noise was deafening), "now we proceed to rub in the lesson with a mild electric shock."

He waved his hand again, and the Head Nurse pressed a second lever. The screaming of the babies suddenly changed its tone. There was something desperate, almost insane, about the sharp spasmodic yelps to which they now gave utterance. Their little bodies twitched and stiffened; their limbs moved jerkily as if to the tug of unseen wires.

"We can electrify that whole strip of floor," bawled the Director in explanation. "But that's enough," he signalled to the nurse.

The explosions ceased, the bells stopped ringing, the shriek of the siren died down from tone to tone into silence. The stiffly twitching bodies relaxed, and what had become the sob and yelp of infant maniacs broadened out once more into a normal howl of ordinary terror.

"Offer them the flowers and the books again."

The nurses obeyed; but at the approach of the roses, at the mere sight of the gaily coloured images of pussy and cock-a-doodle-doo and baa-baa black sheep, the infants shrank away in horror; the volume of their howlings suddenly increased.

"Observe," said the Director triumphantly. "Observe."

Books and loud noises, flowers and electric shocks—already in the infant mind these couples were compromisingly linked; and after two hundred repetitions of the same or a similar lesson would be wedded indissolubly. What man has joined, nature is powerless to put asunder.

"They'll grow up with what the psychologists used to call an 'instinctive' hatred of books and flowers. Reflexes unalterably conditioned. They'll be safe from books and botany all their lives." The Director turned to his nurses. "Take them away again."

Although the experiment described above, and many others, would probably be successful, it seems very rash to describe the mind as wholly built up of conditioned reflexes. Yet this is what some Behaviourists do. They regard the human mind as almost infinitely plastic, so that, by proper conditioning and environment, a baby can be turned into any kind of man. Dr. Watson says that the Behaviourist "believes that, given the relatively simple list of embryological responses, which are fairly uniform in infants, he can build (granting that environment can be controlled) any infant along any specific line—into rich man, poor man, beggarman, thief." This doctrine that, but for the accidents of circumstance, they might have been Einsteins, Shakespeares, or even Henry Fords, is doubtless comforting to self-respecting morons, but to others it will appear merely as another influence of the hasty, even silly, generalizations to which Behaviourists are prone.

The conditioned reflex theory is most successful in its application to language, although even here it does not take us very far. It explains the way in which an infant learns single words in a fairly satisfactory way. Thus a child learns to associate the word "bottle" with the appearance of his food. His learning of such words can be explained as a conditioned reflex. When we come to sentences, however, another element seems to be involved. Differently worded sentences can express the same meaning, and a child is soon able to use sentences, which he has never heard before, to express correctly his meaning. This is obviously not done by any process of trial and error in combining words that he knows. The child first grasps the meaning of the sentence as a whole, and then seizes, from his limited vocabulary, suitable words, and arranges them in a suitable order, for the expression of his meaning. The grasp of the situation as a whole precedes his attempt to express it.

This capacity is not peculiar to man. Certain performances, even of the higher apes, cannot be accounted for by a trial and error process. Kohler's famous experiments on chimpanzees makes this quite clear. He gives several instances of intelligent behaviour on their part which are certainly not instances of trial and error. One of the best known is where the ape Sultan is provided with two bamboo canes with which he tries to draw fruit into his cage. Having found that each of these canes was too short for the purpose, Sultan had a period of silent thought. He found, perhaps accidentally, that one end of one cane could be fitted into one end of the other. Directly he discovered this, he realized that he had the solution of his problem and immediately utilized the lengthened cane he thus obtained to draw in his food. Kohler, speaking of his experiments generally, says:

> It is certainly not a characteristic of a chimpanzee, when he is brought into an experimental situation, to make any chance movements, out of which, among other things, a non-genuine solution could arise. He is very seldom seen to attempt anything which would have to be considered accidental in relation to the situation (excepting, of course, if his interest is turned away from the objective to other things). As long as his efforts are directed to the objective, all distinguishable stages of his behaviour (as with human beings in similar situations) tend to appear as complete attempts at solutions, *none* of which appears as the product of accidentally arrayed

parts. This is true, most of all, of the solution which is finally successful. Certainly it often follows upon a period of perplexity or quiet (often a perod of survey), but in real and convincing cases the solution never appears in a disorder of blind impulses. It is one continuous smooth action, which can be resolved into its parts only by the imagination of the onlooker; in reality they do *not* appear independently. But that in so many "genuine" cases as have been described, their solutions as wholes should have arisen from mere chance, is an entirely inadmissible supposition.

The Behaviourist explanation of other mental processes, such as memory, is similarly deficient. Dr. Watson says: "By 'memory,' then, we mean nothing except the fact that when we meet a similar stimulus again after an absence, we do the old habitual thing (say the old words and show the old visceral-emotional behaviour) that we learned to do when we were in the presence of that stimulus in the first place." Bertrand Russell points out that, if by "old habitual thing" in this description Dr. Watson means verbal habits, then his account is certainly untrue. We can, and do, describe a past experience in many different forms of words, such as "I met Mr. Jones in the train today," and "Joseph was in the 9.35 this morning." The constant element in these different descriptions is their meaning. If they are mere verbal habits then we must assume that we have been spending a large portion of our lives in rehearsing every possible form of words in which everything we can remember can be expressed. Here again we have an example of the exaggeration which so fatally obscures whatever merit the observations of the Behaviourists may possess.

Besides putting forward theories on a basis of fact all too small to support them, the Behaviourists also put forward theories which are opposed to direct experience. They deny obvious facts. Thus they deny that we can form images in our minds—visual, auditory, and so on. Those of us who are strong visualizers, or who take pleasure in mentally "hearing" favourite musical works, know that this statement is merely untrue. Indeed, Behaviourism, when applied to all but very simple mental processes, is so inadequate as to be uninteresting.

§ 4

Behaviourism, it may be said, suffers from the paucity of its initial assumptions; it oversimplifies the mind. We can hardly say the same about that other great group of theories called Psycho-Analysis. The chief interest of psycho-analysis, as a general psychological theory, resides in its assumptions, and of these the chief is its assumption of the existence of what it calls the "Unconscious." We are to suppose that, besides the mental processes of which we are aware, there are mental processes actively going on of which we are quite unaware. Some of these occurrences can be brought into consciousness by a voluntary effort, while others cannot be captured except by the special technique of psycho-analysis or some equivalent method.

That our minds contain elements which are normally inaccessible to us was made clear by the phenomena of hypnotism. By plunging a man into a state of profound hypnosis he could be made to remember events which had long vanished from his normal mind, and which he was quite unable to recover by ordinary voluntary effort—events belonging, for example, to his very early childhood. These memories were only accepted by the investigator, of course, after they had been subsequently verified. On recovering from hypnosis the patient had again lost all memory of these early events and, indeed, of everything that took place during his trance. But if, during hypnosis, the patient was told that he must remember these events in his normal waking condition, it was found that he could do so. Thus memories would be brought into the conscious mind which had long been absent from it.

It is not necessary, in order to account for these results, to suppose that there are literally unconscious memories. We are not forced to conclude that the patient has been unconsciously remembering these early events all through his life. For those psychologists may be right who say that mental events leave "traces" in the brain, and that these traces, in suitable conditions, can give rise to memories. Traces are not themselves supposed to be mental; they may be merely molecular arrangements in certain parts of the brain. We may then suppose that, in hypnosis, the patient's body is thrown into a suitable state for these traces to produce their appropriate mental effects—the memories. Thus this evidence would not, by itself, require us to believe that there are literally unconscious mental events or processes.

The matter is somewhat more complicated when we consider some other phenomena of hypnotism. Thus a man under hypnosis may be told that he cannot bend his arm. Sometimes he says, "I know I can't," and does not even try. At other times he tries without succeeding. But the point is that he is not trying with an undivided will. He has accepted the hypnotist's suggestion, and he intends, "at bottom," as it were, to keep his arm straight. A similar state of affairs in the normal state is given by William James. "Try to feel as if you were crooking your finger, whilst keeping it straight. In a minute it will fairly tingle with the imaginary change of position; yet it will not sensibly move, because *its not really moving* is also a part of what you have in mind." We might suppose, therefore, that the hypnotized person is unable to bend his arm simply because he is playing a game with himself. But as against this, we have the fact that the suggestion can be prolonged into his waking life. He now tries to bend his arm with all his conscious will, and finds that he cannot. If we still suppose that, at bottom, he is willing to keep his arm straight, then this at-bottom-will is something of which he is not conscious. He is unconsciously willing.

Some of the phenomena of hysteria are still more remarkable, and the most striking of these are the cases of double personality. In these cases what seem to be two entirely different persons alternate in their control of the same body. In the cases of what are called co-consciousness one of these persons is aware of the mental processes of the other, although his own mental processes are known only by himself. It may even be that he can fill up gaps in the other personality's conscious processes, that is, that he can report mental events of which the other personality was unconscious. It has even been suggested that in all of us there is another personality, aware of our unconscious as well as of our conscious processes, but that this second personality never, in normal cases, emerges and takes control of the body. However this may be, there is sufficient evidence, from the well-known phenomena of hypnosis and hysteria, to make us willing to lend credence to the hypothesis of an unconscious region of the mind. We cannot take the further step and say that this unconscious region is itself a mind until we find that its structure and operations are of the kind that we attribute to a conscious mind. It may be, in fact, that the unconscious region is too loose and incoherent a collection of mental events and processes to be called a mind at all. The most thorough-going attempts to obtain knowledge of this region have been made by the psycho-analysts.

To begin with, Freud, the founder of psycho-analysis, used hypnotism in his investigation of the unconscious. His teacher, Breuer, had found that hysterical symptoms are in some way related to forgotten events in the past life of the patient. If these events can be restored to consciousness, the symptoms disappear. Breuer therefore took advantage of the increased power of memory conferred by hypnosis to recover buried memories. At first Freud also used this method, but it could not be applied in every case, and Freud determined to dispense with it. It seemed to Freud that the forgotten events had not merely lapsed from the conscious mind, but had been forcibly ejected from it, and he found that there were forces at work in the patient's mind which offered considerable resistance to the recovery of these memories. It occurred to Freud that these two phenomena sprang from the same cause, that the force which had obliterated the forgotten event from the conscious mind was the same force that opposed the recovery of the memory of that event.

Thus originated the important psycho-analytic doctrine of *repression.* If the memory of certain experiences would now cause us pain, these experiences have a tendency to be repressed into the unconscious region of our mind. The unconscious is, indeed, very largely populated by elements whose emergence into the conscious mind would be felt as distressing. Thus, besides painful memories, we have wishes that we should blush to own; these wishes are repressed into the unconscious.

But these wishes are only repressed; they are not annihilated. They still have a life of their own, and their struggle to emerge into consciousness, against the repressing force, may lead to serious conflict. This conflict may be so severe as to give rise to hysterical symptoms. Dreams, according to Freud, are due to struggles of this kind. A dream is, as it were, a disguise under cover of which a forbidden element enters consciousness. These disguises are usually very elaborate. What a dream is really "about" can, it appears, only be discovered by a process of analysis. Dreams which seem perfectly innocuous are shown by the psycho-analyst to be, in reality, extremely complicated and misleading masks for the most revolting contents. Certain hysterical symptoms also are to be interpreted as disguises.

The general doctrine of repression and unconscious wishes applies very well to certain phenomena that became very prominent during the war. Many cases were known of soldiers who suffered from blindness, paralysis, or what not, when there was in fact nothing physically wrong with them at all. The old type of army doctor concluded, of course, that the men were malingering. But it was found, in course of time, that such men were not malingering. Their repressed desire to escape from the war had seized upon some opportunity, provided by some shock, to cause a feigned physical disability in order to achieve its purpose. But this physical disability was by no means a feigned one to the conscious mind. The conscious mind took the feigned disability quite seriously, and only when the repressed desire was resurrected could the disability be made to disappear. A typical case is that of a young officer who fell and bruised his arm. The arm was put into a sling, but, when the sling came to be removed, the arm was found to be rigid and paralysed. There could be no doubt of the officer's *bona-fides* or of the genuineness of his disability. But under hypnosis it became clear that the paralysis had been brought about by the officer's great but repressed fear of returning to duty. This fear, repressed into the unconscious, had ensured its gratification, as it were, by paralysing the arm and so withdrawing the patient from active service. In this case the raising of the repressed impulse to consciousness was attended by much resistance, and provoked passionate outbursts of anger and hatred. Similar instances have been reported of paralysed hands, vocal organs, etc., all devices, on the part of the unconscious, to withdraw the patient from danger. War-time cases of loss of memory also come under this head.

There is considerable evidence, as we have seen, for the existence of the unconscious and for the facts of conflict and repression, although it would be by no means true to say that these doctrines are universally admitted by psychologists. But amongst those who accept these doctrines there is considerable disagreement as to the contents of the unconscious. Freud, as is well known, lays immense stress on repressed sexual desires. This is very apparent in his analysis of dreams.

The ingenuity he displays in giving to a dream a sexual significance is astonishing. But it is, very often, more astonishing than convincing to the normal reader.

The actual desires which are trying to enter consciousness in the dream disguise form what is called the "latent content" of the dream, while the dream as remembered is called the "manifest content." The correct interpretation of a dream, the passage from the manifest content to the latent content is, it appears, a very ticklish business. The masks assumed by the desires which are trying to enter consciousness are very baffling indeed. Thus it may happen, we are told, that a single element in the manifest content stands for several elements in the latent content. A dream which appears to be about one person may, "in reality," be about several people. And the opposite may occur. What appears to be several elements in the manifest content may all boil down to one element in the latent content when the dream is properly interpreted.

Another source of confusion is "displacement," so that a dream which appears to be about one thing is really about something quite different. It may even happen that your dream emotion is the exact opposite of the emotion that inspired the dream. Thus you may wake up weeping over the imagined death of somebody whom, in your unconscious, you wanted to die. The existence of so many possible sources of confusion makes it difficult to believe in any interpretation of any dream. The interpreter has too many variables to play with. It appears evident that, with a little ingenuity, any latent content whatever can be got out of any manifest content.

And, in fact, different analysts would give quite different interpretations of the same dream. Freud, as we know, would find the dream to be a collection of sexual symbols. Adler, who was at one time a psychoanalyst, but who has abandoned Freud's teachings, would find the dream an expression of the Will to Power. Jung would probably give it yet another interpretation. It is quite impossible to say that any one of these interpretations is more plausible than the others. This is distressing if the interpretation of dreams is to be considered a science. It is as if different analytical chemists arrived at quite different constituents for the same chemical compound. It is not surprising, therefore, that this part of Freud's teaching has aroused a great deal of scepticism.

Nevertheless, even an unsympathetic critic must admit that, in some cases, dreams have been plausibly connected with early sexual experiences. The patient, under treatment, could be made to remember and describe these. Freud at first took these descriptions quite seriously, and was duly impressed by the state of social morality revealed by the commonness of these stories of childish seduction, etc. Further experience, however, convinced him that many of these stories are mere fantasies, invented by the patient. But this fact, too, has its own significance, and Freud expanded his general theory to accommodate it. The theory of psycho-analysis is, indeed, a very complicated structure, and it is just this richness of hypothesis that, for many minds, detracts so greatly from its credibility.

A good instance of this fertility in theory-making is provided by Freud's theory of the Libido. The term Libido is thus defined by Freud:

> We call by that name the energy of those instincts which have to do with all that may be comprised under the word "love." The nucleus of what we mean by love naturally consists . . . in sexual love with sexual union as its aim. But we do not separate from this . . . on the one hand, self-love, and on the other, love for parents and children, friendship, and love for humanity in general, and also devotion to concrete objects and to abstract ideas. Our justification lies in the fact that . . . all these tendencies are an expression of the same instinctive activities; in relations between the sexes these instincts force their way towards sexual union, but in other circumstances they are diverted from their aim or are prevented from reaching it, though always preserving enough of their original nature to keep their identity recognizable (as in such features as the longing for proximity and self-sacrifice).

This highly general concept, the libido, is supposed to pass through various stages of development. In the very young infant it is associated with the nutritional process. For, according to Freud, the infant sucking at the mother's breast is, besides satisfying its hunger, also experiencing a sort of sexual pleasure. It is for this reason that it likes sucking its thumb or other non-nutritious object. Later on in the child's history the libido becomes associated with the act of excretion, and still later on, if all goes well, it is transferred to the organs of generation. Thus, what we call the sexual instinct is not something new that emerges for the first time at a certain age; it is merely a different bodily localization of something that has been present from the beginning. But besides the change in bodily localization the libido also undergoes development with respect to the objects towards which it is directed. At first the object of the libido is the child's own body. This is the "Narcissistic" stage. Later on the parents, and then brothers and sisters, become the objects towards which the libido is directed. Further development, if normal, transfers the libido to objects in the outer world, outside the family.

In many cases, however, the development of the libido has not been normal. Any one of the libidinal impulses may be arrested at some point in this process. Such an arrest is called a "fixation." A diminished volume of libido is then available for future development, and this diminished energy is so much the less likely to achieve satisfaction in adult life. It is very hard, in any case, in modern social and cultural conditions, to achieve as full a love-life as one would desire. Most people encounter "privation"; they achieve but a limited indulgence of their sexual desires. And when the libidinal energy has been diminished by fixation the resulting conflict may have serious consequences. The thwarted libidinal energy undergoes "regression," it flows back to the points of fixation, and so revives desires and impulses appropriate to an earlier age. Freud illustrates the idea by an analogy:

> If you think of a migrating people who have left large numbers at the stopping-places on their way, you will see that the foremost will actually fall back upon these positions when they are defeated or when they meet with an enemy too strong for them. And, again, the more of their number they leave behind in their progress, the sooner will they be in danger of defeat.

In an individual who does not permit these revived childish desires to achieve satisfaction the libidinal impulse may be driven right back into the unconscious and, working there, provoke hysterical symptoms and thus achieve a sort of satisfaction in a disguised form. In people who do not develop a neurosis in consequence of privation, the libido appears to be more mobile, as it were, and can be diverted to substitute objects. The most important form of substitution is "sublimation" whereby art, science, social reform, or what not, becomes the object of the thwarted libido. But sublimation, we are told, can never carry off the whole of the libido, and those people who have achieved it are often unhappy, dissatisfied, eccentric.

This libido theory is very representative of the fertile imagination that has built up psycho-analysis. It is a plausible, fluent structure, and yet the facts on which it is built admit of entirely different interpretations. This is sufficiently proved by the existence of rival systems, of which those of Jung and Adler are probably the best known. Jung also postulates something he calls libido, but his libido is quite different from that of Freud. According to Jung the libido is an undifferentiated primal life-force from which all the instincts derive. In the infant it assumes the form of the instinct of nutrition. It is only much later that it assumes the sexual form. Then he denies that the unconscious is a region inhabited by desires which have been repressed after conflict. The unconscious, according to Jung, arises as a consequence of the individual's one-sided mental growth. Jung, besides many subdivisions, divides people into two main classes, extroverts and introverts. In extroverts feeling is more developed than thought; in introverts thought is more developed than feeling. In each case the neglected potentialities tend to become unconscious. If, now, the introvert is faced by a situation that demands feeling rather than thought, a conflict is set up in him which may lead to a neurosis. And the same thing may happen to the extrovert when more thought than feeling is required. Jung, therefore, does not look to the patient's past life for the causes of his neurosis. Rather, he endeavours to discover what the present task is that the patient is shirking, and then tries to elicit those buried resources in his unconscious that will enable him to deal with it. And naturally, with this outlook, Jung gives an entirely different interpretation of dreams. He believes that dreams reveal the attitude of the unconscious towards the tasks of life.

Adler's interpretation is again different, but is equally plausible. Adler sees the driving force of life as the urge to acquire power and superiority over one's fellows. This outlook enables Adler to give analyses which are quite as convincing and penetrating as those that follow from entirely different theories. Thus we all know the invalid who, thwarted in his desire to obtain power and superiority by normal means, uses his very disability to dominate and tyrannize over the whole household. Where Freud finds a sexual content in a neurotic fantasy, Adler has no difficulty in finding yet another illustration of his will-to-power theory. Besides rejecting Freud's libido, Adler also ignores the unconscious and makes no use of the notion of repression. Thus, although he started as a psycho-analyst, Adler no longer has any resemblance to one, since he has rejected the very foundations of Freud's theory.

Other break-aways from Freud's doctrines, more or less thorough, could be enumerated. In fact, psycho-analysis bids fair to rival Christianity in the number of its sects. And, like the Christian sects, all forms of psycho-analysis boast a comprehensive and adequate outlook, and can point to an impressive list of spiritual and bodily cures in proof of the efficacy of their teachings. As a science, therefore, any form of psycho-analysis is in an unsatisfactory position. It cannot appeal to results as a proof of its theory, for those results, the cures it effects, can be obtained equally well by an entirely different theory. The theories therefore are not to be judged by the results. It follows that they can only be judged by their initial probability. And here we encounter the difficulty that nearly all forms of psycho-analysis sound equally probable. Perhaps a difference should be made here in favour of those theories which accept an unconscious as against those that do not. The concept of the unconscious is certainly important, but it existed long before psycho-analysis was thought of. So far as psycho-analysis has introduced anything new, it is debatable.

There are other psychological theories that could be discussed, such as the *Gestalt* theory to which we referred earlier, but there is nothing in these theories that seriously modifies our conviction that psychology is not yet a science. Other studies, such as sociology, economics, and so on, also have, from the scientific point of view, something unsatisfactory about them. Science is at its strongest in dealing with the material universe. Its pronouncements on other matters are relatively weak and hesitating.

CHAPTER VI

The Limitations of Science

We have seen that the scientific account of our universe appears clearest and most convincing when it deals with inanimate matter. Here we feel that the account is relatively satisfactory because it does, on the whole, meet the kind of interest we take in these phenomena. The age, position, size, velocity, chemical constitution of a star is, for instance, the kind of information we want about a star. And to be told that matter consists of little electrified particles arranged with respect to one another in certain ways makes us feel that our curiosity about matter has very largely been met.

But when we come to the sciences dealing with life, the state of affairs is less satisfactory. Many of the questions that seem to us quite fundamental have not been met. What, for instance, makes us regard a living organism as a whole, and not merely as the sum of its parts? What does this vague notion of "wholeness" or "individuality" really amount to? Even if every bodily activity of the animal was explained in terms of physical and chemical changes, we should still feel that our question was unanswered unless what appears as the *purposive order* of those changes was also accounted for. But "purpose" is not yet a scientific notion. It is not employed in the physical and chemical sciences, and the majority of biologists or, at least, of physiologists, are reluctant to introduce any ideas which have not been found necessary in these sciences. This is doubtless an excellent procedure so far as certain limited classes of problems are concerned, but it also seems to lead to the consequence that the most obvious and fundamental problems of biology are not even approached. The case has been put forcibly but fairly, by Professor Whitehead. Speaking of the application of the notions of physics and chemistry to life he says:

> The brilliant success of this method is admitted. But you cannot limit a problem by reason of a method of attack. The problem is to understand the operations of an animal body. There is clear evidence that certain operations of certain animal bodies depend upon the foresight of an end and the purpose to attain it. It is no solution of the problem to ignore this evidence because other operations have been explained in terms of physical and chemical laws. The existence of a problem is not even acknowledged. It is vehemently denied. Many a scientist has patiently designed experiments for the *purpose* of substantiating his belief that animal operations are motivated by no purposes. He has perhaps spent his spare time in writing articles to prove that human beings are as other animals so that "purpose" is a category irrelevant for the explanation of their bodily activities,

> his own activities included. Scientists animated by the purpose of proving that they are purposeless constitute an interesting subject for study.
>
> Another reason for the extrusion of final causation is that it introduces a dangerous mode of facile explanation. This is certainly true. The laborious work of tracing the sequence in physical antecedents is apt to be damaged by the facile suggestion of a final cause. Yet the mere fact that the introduction of the notion of final causation has its dangers is no reason for ignoring a real problem. Even if heads be weak, the problem remains.

Whitehead's indictment certainly seems to be justified. Again and again we have the feeling that the primary concepts used by biologists are inadequate to their most important problems. The great theory of natural selection, for example, when studied in detail, is full of lacunæ. Instead of the natural and ready assent one gives to a demonstration in physics, for example, one has to make a really immense effort to believe, even for a moment, that the whole evolution of living forms on this planet has come about by "random variations" and the struggle for existence. It does not in the least explain the most obvious fact about the whole process, that is, the upward tendency of living things. If mere survival is the sole desideratum, then it would seem that some rudimentary type of organism would be all that is needed. And there would seem no reason why even a rudimentary type of organism should appear, since it could not hope to rival in longevity the everlasting rocks.

We have the impression that it is only by an extraordinary act of faith that biologists can suppose that the actual progress of life can be explained in the terms they adopt. There are many biologists, of course, who deny that such an explanation is possible. They have introduced such notions as "vital force," "entelechy," "notism," and so on, but they have not succeeded in making such terms definite enough for scientific purposes. But such terms testify to a conviction that the present primary concepts of biology are inadequate.

We find a similar groping after some new conception in a speculation thrown out by one of the most brilliant of the younger biologists, J.B.S. Haldane:

> I imagine that associated with an evolving line there may be some "emergent," just as mind is associated with brain. Royce (1901) tried to give a concrete picture of such an emergent as a mind with a vast time-scale, and suggested that the intense feelings associated with reproduction were in that mind as well as our own. If there is an element of truth in such speculations, I question whether such an emergent should be regarded as probably mind-like. . . . My suspicion of some unknown type of being associated with evolution is my tribute to its beauty, and to that inexhaustible queerness which is the main characteristic of the universe that has impressed itself on my mind during twenty-five years of scientific work.

Psychology, contrasted with the physical sciences, seems even more inadequate. The psychological theory that sticks most closely to the terms of mechanics is Behaviourism, and we have seen that that theory is hopelessly inadequate. Psycho-analysis, it is true, introduces primary concepts which are not mechanical, but these concepts are far too vague and indefinite to be called scientific. Such a concept as Freud's libido, for example, is called upon to explain so much that it explains nothing. So far as a scientific explanation is concerned, no more is gained by saying that the most amazingly diverse manifestations all come about through the libido than by saying that they all come about through the Will of God. The explanation admittedly explains everything—and therefore explains nothing.

We conclude, therefore, that it is only in the sciences dealing with inanimate matter that a reasonably adequate set of primary concepts has yet been isolated. We say "reasonably adequate" because in relativity and quantum theory, even these concepts are undergoing extensive revision. Nevertheless, they have been a tremendous success over three centuries, and it is doubtless owing to this success that scientific men have been led to adopt them in fields where they would not, perhaps, naturally suggest themselves. What are these concepts, and how were they isolated? The answer to this question will enable us to realize the nature and limitations of the scientific method, and also show us how baseless is the hope that these concepts will ever suffice to cover the whole of experience.

§ 2

Scientific method, as we see from the work of its founders, Copernicus, Kepler, Galileo, began by quite consciously and deliberately selecting and abstracting from the total elements of our experience. From the total wealth of impressions received from nature these men fastened upon some only as being suitable for scientific formulation. These were those elements that possess *quantitative* aspects. Between these elements mathematical relations exist, and these men were convinced that mathematics is the key to the universe. It is interesting, in view of its immense importance, to know how they came by this belief, for it was by no means a general persuasion. It was not part of the dominant Aristoteleanism of their time. It seems to have been in part due to an innate prejudice, very proper to born mathematicians, and in part to the neo-Platonic philosophy that was prevalent, at that time, in South Europe. This philosophy contained important Pythagorean elements, and gave to the mathematical aspects of the universe a much more exalted position than they occupied in the current Aristotelean outlook. Copernicus became acquainted with this philosophy during his stay in Italy, and also with the fact that some of the ancient Greek philosophers had put forward the hypothesis that the earth was in motion. Being led, in this way, to take the sun as his centre of reference, Copernicus found, as we have already said, that a great harmony was bestowed on the motions of the heavenly bodies. With the incorrect ideas of dynamics prevalent at the time, Copernicus's theory was open to grave objections. As a physical explanation of phenomena it was certainly no better than the Ptolemaic theory. Nevertheless, Copernicus was confident that its superior æsthetic charm would be sufficient to commend it to mathematicians.

This expectation was justified. We have already given a quotation from Kepler which shows how greatly the æsthetic charm of the new theory appealed to him. To the mind of Kepler, however, the claims of the theory were greatly reinforced by the dignified position it gave to the sun, for Kepler, in a vague and mystical fashion, was a sun-worshipper. This very unscientific attitude may be illustrated by a quotation from what was probably an early lecture:

> In the first place, lest perchance a blind man might deny it to you, of all the bodies in the universe the most excellent is the sun, whose whole essence is nothing else than the purest light, than which there is no greater star; which singly and alone is the producer, conserver, and warmer of all things; it is a fountain of light, rich in fruitful heat, most fair, limpid, and pure to the sight, the source of vision, portrayer of all colours, though himself empty of colour, called king of the planets for his motion,

> heart of the world for his power, its eye for his beauty, and which alone we should judge worthy of the Most High God, should he be pleased with a material domicile and choose a place in which to dwell with the blessed angels. . . . Since, therefore, it does not befit the first mover to be diffused throughout an orbit, but rather to proceed from one certain principle, and, as it were, point, no part of the world, and no star, accounts itself worthy of such a great honour; hence by the highest right we return to the sun, who alone appears, by virtue of his dignity and power, suited for this motive duty and worthy to become the home of God Himself, not to say the first mover.

Thus Kepler's acceptance of the Copernican theory was very largely due to motives that we should now regard as quite unscientific. Kepler had a preconceived idea as to the sort of thing the universe is. He did not approach the facts with the docility and lack of prejudice that is proper to the ideal scientific investigator. His deepest conviction was that nature is essentially mathematical, and all his scientific life was an endeavour to discover nature's hidden mathematical harmonies. Galileo, also, had no doubt that mathematics is the one true key to natural phenomena. It was this persuasion that gave these men their criterion for selection amongst the total elements of their experience. They confined their attention to those elements amongst which mathematical relations exist. Bodies, for instance, have for their measurable aspects size, shape, weight, motion.

Such other characteristics as they possess were regarded as belonging to a lower order of reality. The real world is the world of mathematical characteristics. In fact, our minds are so constructed, Kepler said, that they can know nothing perfectly except quantities.

With Galileo this separation of the mathematical from the other qualities became a perfectly clear and definite doctrine. Kepler had supposed that the non-mathematical qualities actually did belong to bodies, but that they were somehow less real. Galileo went further than this, and stated that the non-mathematical properties are all entirely subjective. They have no existence at all apart from our senses. Thus colours, sounds, odours, and so on exist, as such, wholly in our minds. They are, in reality, motions of some kind or another in the external world, and these motions, impinging on our senses, give rise to these sensations of colour, sound, and so on. It is the mind that peoples the world with the songs of birds, the colours of the sunset, etc. In the absence of mind the universe would be a collection of masses of various sizes, shapes, and weights, drifting, without colour, sound, or odour, through space and time. As the doctrine has turned out to be so immensely important, we may give Galileo's own exposition of it. After explaining that what we perceive as heat is really a species of motion in the heated body, he goes on:

> But first I want to propose some examination of that which we call heat, whose generally accepted

notion comes very far from the truth if my serious doubts be correct, inasmuch as it is supposed to be a true accident, affection, and quality really residing in the thing which we perceive to be heated. Nevertheless, I say, that indeed I feel myself impelled by the necessity, as soon as I conceive a piece of matter or corporeal substance, of conceiving that in its own nature it is bounded and figured in such and such a figure, that in relation to others it is large or small, that it is in this or that place, in this or that time, that it is in motion or remains at rest, that it touches or does not touch another body, that it is single, few, or many; in short by no imagination can a body be separated from such conditions: but that it must be white or red, bitter or sweet, sounding or mute, of a pleasant or unpleasant odour, I do not perceive my mind forced to acknowledge it necessarily accompanied by such conditions; so that if the senses were not the escorts, perhaps the reason or the imagination by itself would never have arrived at them. Hence I think that these tastes, odours, colours, etc., on the side of the object in which they seem to exist, are nothing else than mere names, but hold their residence solely in the sensitive body; so that if the animal were removed, every such quality would be abolished and annihilated. Nevertheless, as soon as we have imposed names on them, particular and different from those of the other primary and real accidents, we induce ourselves to believe that they also exist just as truly and really as the latter.

I think that by an illustration I can explain my

meaning more clearly. I pass a hand first over a marble statue, then over a living man. Concerning all the effects which come from the hand itself, they are the same whether on the one or on the other object—that is, their primary accidents, namely motion and touch (for we call them by no other names)—but the animate body which suffers that operation feels various affections according to the different parts touched, and if the sole of the foot, the knee-cap, or the armpit be touched, it perceives besides the common sense of touch, another affection, to which we have given a particular name, calling it tickling. Now this affection is all ours, and does not belong to the hand at all. And it seems to me that they would greatly err who should say that the hand, besides motion and touch, possessed in itself another faculty different from these, namely the tickling faculty; so that tickling would be an accident that exists in it. A piece of paper, or a feather, lightly rubbed on whatever part of your body you wish, performs, as regards itself, everywhere the same operation, that is, movement and touch; but in us, if touched between the eyes, or on the nose, and under the nostrils, it excites an almost intolerable tickling, though elsewhere it can hardly be felt at all. Now this tickling is all in us, and not in the feather, and if the animate and sensitive body be removed, it is nothing more than a mere name. Of precisely a similar and not greater existence do I believe these various qualities to be possessed, which are attributed to natural bodies, such as tastes, odours, colours, and others.

We are not here concerned to discuss the validity of this division into primary and secondary qualities, but merely to show the role it played in the formation of the scientific outlook. We may mention, however, that it has been attacked, philosophically, in two ways. It has been denied, on the one hand, that the primary qualities are any less "subjective" than the secondary qualities. Both sets of qualities, it has been argued, depend on the mind, and cannot be conceived as existing apart from the mind. On the other hand, it has been denied that the secondary qualities are any less "objective" than the primary qualities. Each of these arguments denies that primary qualities are any more "real" than secondary qualities, although it may be true that only the primary qualities are susceptible of mathematical treatment. But the Galilean doctrine, which became part of the general scientific outlook, maintains that the primary qualities alone are real—the others are in some sense illusory.

With this reduction of the real world to colourless, soundless, odourless bodies in motion, the notions of space and time underwent a profound change. To the medieval philosophers the temporal process was the transformation of potentiality into actuality. The purpose of everything was to reach a higher state of being, culminating in union with God. The whole of the past, up to the present moment, was the ground already won, as it were. As the process continues, the ground won increases; the future is being drawn into the present. This process goes on until the final culmination is reached, when time stands still. We see how different is this notion of time from the mathematical time introduced by Galileo. Time, as it appears in science, may be likened to an ever-moving mathematical point. The present moment, which has no finite duration, is merely a boundary point between a vanished past and a non-existent future. Time, conceived in this way, can be represented very simply mathematically as a straight line, successive points on the line representing successive instants of time. This conception has hitherto proved quite adequate for the needs of science, and has therefore been generally adopted. Philosophically, however, it has brought in its train a host of unsolved puzzles, and there can be little doubt that a new conception of time will have to be formulated. Even science itself is now giving hints that this may be necessary.

With this change in the notion of time comes a corresponding change in the notion of cause. When all things were regarded as moving towards union with God, then union with God was regarded as the final cause of all change. The cause of a process was to be found in the end towards which it tended. The reason why things happened was to be found in the purpose the happening served. With the new notion of time the future, being non-existent, had no influence on present happenings. The cause of anything happening now was to be found in its immediate past. Further, all that really happens are motions—motions of the constituent particles of the bodies forming the real world. And these motions are themselves the products of preceding motions. Galileo also believed that there are "forces," which are revealed by motions, but as to the nature of force he refrained from speculating.

We see that the scientific outlook, as presented by Galileo, constitutes a really amazing revolution in thought. The vivid world of the medievalist, a world shot through with beauty and instinct with purpose, a world having an intimate relation to his own destiny and an intelligible reason for existing in the light of that destiny, is dismissed as an illusion. It has no objective existence. The real world, as revealed by science, is a world of material particles moving, in accordance with mathematical laws, through space and time.

The early creators of science did not assume that this real world was also purposeless. Although God was no longer invoked as the final cause of phenomena, He was still given an important role as the initiator of the whole process. But, having made matter with its properties, He then left the world to develop according to the laws of mathematical necessity. In course of time it came to be considered unnecessary to invoke God even for this purpose, and the way was clear for thorough-going materialism. That this doctrine was made plausible by scientific researches is the greatest possible testimonial to the "humanistic" importance of science. We have seen what the medieval outlook was in essentials. The essentials of the materialistic outlook may be given in the words of Bertrand Russell:

> That man is the product of causes which had no prevision of the end they were achieving; that his origin, his growth, his hopes and fears, his loves and beliefs, are but the outcome of accidental collocations of atoms; that no fire, no heroism, no intensity of thought and feeling, can preserve an individual life beyond the grave; that all the labours of the ages, all the devotion, all the inspiration, all the noonday brightness of human genius, are destined to extinction in the vast death of the solar system, and that the whole temple of Man's achievement must inevitably be buried beneath the debris of a universe in ruins—all these things, if not quite beyond dispute, are yet so nearly certain, that no philosophy which rejects them can hope to stand.

This astonishing change in outlook has been brought about by assuming that, of all the elements of our total experience, only those elements which acquaint us with the quantitative aspects of material phenomena are concerned with the real world. They alone refer to an objective world. None of the other elements of our experience, our perceptions of colour, etc., our response to beauty, our sense of mystic communion with God, has an objective counterpart. All these things, which are ultimately products of the motions of little particles, are illusory in the sense that they do not acquaint us with the nature of objective reality. Our knowledge of the little particles, itself, of course, a product of their motions, is the only result of their activities which is not illusory in this sense. But the materialistic doctrine had not reached this baffling stage of development in the early days of science. Secondary qualities had been classed as subjective, but men continued to argue as if God and His purposes were still part of reality.

A further step towards separating the quantitative from the other features of perception was made by Descartes. He regarded extension as the one essential property of material bodies, and he invented an ether to whirl them about in, in order to account for such properties as weight, velocity, etc. This part of Descartes's work, which was subjected to a deadly analysis by Newton, is of no scientific importance. But what is of importance is the division he made between this extended world and the unextended world of mind. It was quite clear to Descartes that man has a mind. *Cogito, ergo sum.* But this mind he regarded as completely unextended—the notion of spatial magnitude could not be applied to it. It was in this mind that all the secondary qualities existed—indeed all qualities except the quality of extension. The division between mind and matter was made complete. It became a puzzle, therefore, to know how matter could act on such a mind, producing feelings, sensations, etc., and how the mind could know anything of such a material world. Descartes could say no more than that God had made this interaction possible.

Descartes carried the theory that the real world is a world of mathematical characteristics as far as it could go. In fact, he taught that the only properties inherent in matter itself are geometrical properties. Other properties, such as mass and weight, were vaguely accounted for by an imaginary "ether." His doctrine that the mind, itself non-spatial, influenced the body through the brain, was popularized to mean that the mind is located in some small region of the brain. Amongst the English writers Hobbes objected to this dualism, and proposed to abolish the mind as an independent reality, maintaining that everything, thought included, was produced by bodies in motion. More, a writer who influenced Newton, made the different objection that nothing that exists can be conceived as non-extended. He regarded the mind as able to fill the body and even, on occasion, to protrude from it as a sort of effluvium all round it. The mind of God, in particular, is to be regarded as infinitely extended throughout space and time. But all these ideas may be regarded as more philosophical than scientific. Their influence in forming the scientific outlook was only indirect, and it was not permanent.

§ 3

We have seen that, so far, the developing scientific outlook owed its main features to the predilections of the mathematician. And the main assumption of the philosophy accompanying this scientific achievement is that the real may be identified with the quantitative. Compared with the fully developed modern scientific outlook, we see that these early men of science were too prone to legislate for the universe on the basis of certain *a priori* assumptions—assumptions which were really expressions of their dominating mathematical predilections. The modern outlook differs from theirs by the more tentative character of its assumptions. Although mathematics has hitherto proved itself by far the most powerful tool for the scientific investigation of nature, it is no longer an article of belief that nature is necessarily mathematical. We now realize that mathematical deductions, however rigorous, must always be checked by experiment. Even Galileo, who strikes us as the most modern of all these early workers, tells us that often he did not consider experimental checks on his mathematical reasoning to be necessary, and undertook the experiments solely in order to convince his opponents. The first perfect fusion of the mathematical with the empirical outlook was accomplished by Isaac Newton.

Newton was not the first, of course, to realize the need of experiment. All his predecessors had realized this to some extent, and such men as Gilbert, Harvey, Boyle, were complete empiricists. But Newton was the first who ideally combined, in his own method, the two ways of approach. This combination was so unusual that none of his contemporaries understood it. It is still unusual in practice, although recognized as the ideal scientific method in theory. Mathematicians are still prone to have what the experimentalists consider an excessive confidence in their mathematical deductions, and the experimentalist sometimes seems to the mathematician to display an unnecessary degree of caution. Newton not only combined the attitude of both types; he also combined their powers. He was not only a supreme mathematician; he was also a great experimentalist.

Newton starts with no *a priori* ideas about the constitution of the world. He does not regard the world as necessarily mathematical, although the successes achieved by his predecessors show that mathematics is a very hopeful tool to employ. He tells us, in the *Principia,* that he has adopted the mathematical method of approach, and that it has led him to certain principles. But, unlike his great predecessors, he does not assume that mathematics is the one key to reality. He says: "I hope the principles here laid down will afford some light either to that or some truer method of philosophy." Having decided, in view of its proved efficacy, to employ the mathematical method, he employs it with certain safeguards. He says: "All the difficulty of philosophy seems to consist in this—from the phenomena of motions to investigate the forces of nature, and then from these forces to demonstrate the other phenomena." Newton's method sounds to us nowadays perfectly obvious. He starts by investigating experimentally a simple group of phenomena and disentangling from them their quantitative relations. He then deduces the mathematical consequences of these relations. These deductions, if correct, will cover the behaviour of phenomena over a wider field. Newton accordingly brings these phenomena to the experimental test, to see if his deductions are confirmed. He has no *a priori* standpoint from which he legislates for nature. Experiment is the final test. A fuller exposition of his method may be given in his own words:

> As in mathematics, so in natural philosophy, the investigation of difficult things by the method of analysis, ought ever to precede the method of composition. This analysis consists in making experiments and observations, and in drawing general conclusions from them by induction, and admitting of no objections against the conclusions, but such as are taken from experiments or other certain truths. For hypotheses are not to be regarded in experimental philosophy. And although the arguing from experiments and observation by induction be no demonstration of general conclusions; yet it is the best way of arguing which the nature of things admits of, and may be looked upon as so much the stronger, by how much the induction is more general. And if no exceptions occur from phenomena, the conclusion may be pronounced generally. But if at any time afterwards any exception shall occur from experiments, it may then begin to be pronounced with such exceptions as occur. By this way of analysis, we may

> proceed from compounds to ingredients, and from motions to the forces producing them; and in general, from effects to their causes, and from particular causes to more general ones, till the argument end in the most general. This is the method of analysis; and the synthesis consists in assuming the causes discovered, and established as principles, and by them explaining the phenomena proceeding from them, and proving the explanations.

By "proving the explanations" Newton means, of course, submitting them to the experimental test.

Although Newton did not share the mathematical *a priorism* of his predecessors, he shared their other assumptions. Like them, he dispensed with final causes, and found the cause of a phenomenon in its immediately preceding physical conditions. Also, he proceeded as if science formed a self-enclosed system, that is to say, as if a complete account of phenomena could be given in the terms mass, velocity, force, etc., which science had isolated, and without bringing in such concepts as beauty, purpose, etc., which did not form part of the scientific outlook. It is not possible to say that Newton held these opinions dogmatically, since he himself said that science, in the mathematical form it had assumed, was an adventure, which might have to be replaced by a truer method. But for practical working purposes he certainly made these assumptions, and his immense success caused these assumptions to be unquestioningly accepted by the whole scientific world. Newton also accepted the doctrine of the subjectivity of secondary qualities. He says:

> If at any time I speak of light and rays as coloured or endued with colours, I would be understood to speak not philosophically and properly, but grossly, and according to such conceptions as vulgar people in seeing all these experiments would be apt to frame. For the rays to speak properly are not coloured. In them there is nothing else than a certain power and disposition to stir up a sensation of this or that colour. For as sound in a bell or musical string or other sounding body, is nothing but a trembling motion, and in air nothing but that motion propagated from the object, and in the sensorium 'tis a sense of that motion under the form of sound; so colours in the object are nothing but a disposition to reflect this or that sort of rays more copiously than the rest; in the rays they are nothing but their dispositions to propagate this or that motion into the sensorium, and in the sensorium they are sensations of those motions under the forms of colours.

Newton also accepted the idea that the mind is located somewhere in the brain—the sensorium. But he did not, as some of his followers did, regard man as a sort of irrelevant and accidental product of the vast objective mathematical machine that is external nature. Newton believed that nature was designed by God, and he thought that there were countless evidences to prove it. Further, he thought that God every now and then made adjustments to the mathematical machine, in order to keep it going. These theological ideas of Newton's are of no great interest to us, for they did not long survive. Men came to see that it was not logically necessary to postulate a "First Cause," nor did they find it necessary to suppose that the universe must keep going for ever. Even if it be true, as Newton supposed, that the universe cannot keep going without divine intervention, we have only to accept the alternative, which scientific men are quite willing to do, that the universe will not keep going.

All this part of Newton's thinking dropped away. What remained was his method, and his fundamental concepts, mass, force, etc. These concepts, as he defined them, came to be regarded as ultimate and exhaustive. It was believed that every phenomenon would prove to be explicable in terms of these concepts. This was, in essentials, the outlook of science for nearly two hundred years, and this assumption still profoundly colours scientific thinking.

§ 4

What is called the modern "revolution" in science consists in the fact that the Newtonian outlook, which dominated the scientific world for nearly two hundred years, has been found insufficient. It is in process of being replaced by a different outlook, and, although the reconstruction is by no means complete, it is already apparent that the philosophical implications of the new outlook are very different from those of the old one.

Science has become self-conscious and comparatively humble. We are no longer taught that the scientific method of approach is the only valid method of acquiring knowledge about reality. Eminent men of science are insisting, with what seems a strange enthusiasm, on the fact that science gives us but a partial knowledge of reality, and we are no longer required to regard as illusory everything that science finds itself able to ignore. But the enthusiasm with which some men of science preach that science has limitations is not really surprising. For the universe of science, if accepted as the final reality, made of man an entirely accidental by-product of a huge mindless, purposeless, mathematical machine. And there are men of science sufficiently human to find such a conclusion disconcerting. Even the sturdy Victorians who preached the doctrine betray at times a despairing wish that things were not so. We need not be surprised, therefore, to find that the discovery that science no longer compels us to believe in our own essential futility is greeted with acclamation, even by some scientific men.

This change in the scientific outlook seems to have taken place suddenly. It is not yet sixty years since Tyndall, in his Belfast Address, claimed that science alone was competent to deal with all man's major problems, and it is not yet twenty years since Bertrand Russell, contemplating the scientific answers, said that "only on the firm foundation of unyielding despair can the soul's habitation henceforth be safely built." But, in truth, so far as these remarks sprang from the conviction that the sole reality is "matter and motion," their foundations had already been undermined. The attempt to represent nature in terms of matter and motion was already breaking down. That attempt was at its most triumphant by the end of the eighteenth century, when Laplace was emboldened to affirm that a sufficiently great mathematician, given the distribution of the particles in the primitive nebula, could predict the whole future history of the world. The fundamental concepts isolated by Newton had proved themselves so adequate in the applications that had been made of them that they were regarded as the key to—everything.

The first indication that the Newtonian concepts were not all-sufficient came when men tried to fashion a mechanical theory of light. This endeavour led to the creation of the ether, the most unsatisfactory and wasteful product of human ingenuity that science has to show. For generations this monster was elaborated. Miracles of mathematical ingenuity were performed in the attempt to account for the properties of light in terms of the Newtonian concepts. The difficulties became ever more heartbreaking until, after the publication of Maxwell's demonstration that light is an electromagnetic phenomenon, they seemed to become insuperable. But the ether, by this time, had become too complicated to be credible. It was not only complicated; it was ugly, and ugliness in scientific theories is a thing no scientific man will tolerate if he can possibly help it. Copernicus was a true judge of the scientific temperament when he showed himself confident that the æsthetic charm of his theory would suffice to enable it to make its way against the insufferably complicated theory of Ptolemy. The construction of ethers became a decaying industry, and largely because there was so little demand for the product. For it had dawned on men of science that there was, after all, nothing sacrosanct about the Newtonian entities. It might be that his list of ultimates, mass, force, and so on, was not exhaustive. Instead of reducing electricity to these terms, it might be better to add it to the list. This was done. After a certain amount of hesitation, and a few last desperate efforts to make electricity mechanical, electricity was added to the list of irreducible elements.

This may seem to have been a simple and obvious step to take, but it was, in reality, of profound significance. For the Newtonian concepts were all of a kind that one seemed to understand intimately. Thus the mass of a body was the quantity of matter in it. Inertia was that familiar property of matter which makes it offer resistance to a push. Force was a notion derived from our experience of muscular effort. Of course, all these concepts, in order to be of use to science, had to be given quantitative expression. They entered our calculations as mathematical symbols. Nevertheless, we supposed that we knew the nature of what we were talking about. But in the case of electricity its nature is precisely what we did not know. Attempts to represent it in familiar terms—as a condition of strain in the ether, or what not—had been given up. All that we knew about electricity was the way it affected our measuring instruments. The precise description of this behaviour gave us the mathematical specification of electricity and this, in truth, was all we knew about it.

It is only now, in retrospect, that we can see how very significant a step this was. An entity had been admitted into physics of which we knew nothing but its mathematical structure. Since then other entities have been admitted on the same terms, and it is found that they play precisely the same role in the formation of scientific theories as do the old entities. It has become evident that, so far as the science of physics is concerned, we do not require to know the nature of the entities we discuss, but only their mathematical structure. And, in truth, that is all we do know. It is now realized that this is all the scientific knowledge we have even of the familiar Newtonian entities. Our persuasion that we knew them in some exceptionally intimate manner was an illusion. So far as the science of physics is concerned, the old entities and the new are on the same footing—the only aspects of them with which we are concerned are their mathematical aspects.

With this realization it is no long step to Eddington's position that a knowledge of mathematical structure is the only knowledge that the science of physics can give us. Of all the philosophical speculations which have been hung on to the new physics, this seems to be the most illuminating and the best-founded. It seems to be true that "exact" science is a knowledge of what Eddington calls "pointer-reading"—the readings on an instrument of some kind. We assume, of course, that these readings refer to various qualities of the external world, but all we actually know about these qualities, for the purposes of exact science, is the way they affect our measuring instruments. As Eddington [1] says:

> Leaving out all æsthetic, ethical, or spiritual aspects of our environment, we are faced with qualities such as massiveness, substantiality, extension, duration, which are supposed to belong to the domain of physics. In a sense they do belong; but physics is not in a position to handle them directly. The essence of their nature is inscrutable; we may use mental pictures to aid calculations, but no image in the mind can be a replica of that which is not in the mind. And so in its actual procedure physics studies not these inscrutable qualities, but pointer-readings which we can observe. The readings, it is true, reflect the fluctuations of the world-qualities; but our exact knowledge is of the readings, not of the qualities.

---

[1] "The Domain of Physical Science," Essay in *Science, Religion and Reality*.

The former have as much resemblance to the latter as a telephone number has to a subscriber.

Eddington [1] has given a very remarkable illustration of this point of view in his extension, to which we have already referred, of Einstein's relativity theory. Eddington starts with undefinable "point-events." We do not know, and we do not require to know, what point-events *are*. All we know is that it takes four numbers to indicate a point-event uniquely—as it takes two numbers, the number of the street and the number of the house, to indicate a house in New York uniquely. Between neighbouring pairs of point-events is a relation called the "interval." Again we do not know the nature of this relation, but it has a mathematical aspect, and this is all we require to know. From this primitive material Eddington manages to derive, by mathematical analysis, the world of physics, matter, and its laws. It is true that the derivation is not complete. The atomic constitution of matter—all the quantum phenomena—are left on one side. But all the field phenomena of physics, the conservation of mass and energy, the laws of gravitation and electricity, emerge as necessary consequences of the elementary structural relations assumed in the beginning. Whether or not this derivation of field physics—we know that Einstein prefers a different derivation—is of permanent scientific importance, it is obviously a very convincing

[1] *The Mathematical Theory of Relativity.*

illustration of Eddington's contention that exact science deals wholly with structure. For his derivation is not affected whatever nature we attribute to the point-events that constitute all phenomena. It could even be maintained that they are happenings in the mind of God if, like Newton, we supposed space to be God's "sensorium." Science throws no light on such questions. It would be confined to remarking, of the Big Four at Versailles, that they numbered four, to use one of Eddington's illustrations.

The fact that science is confined to a knowledge of structure is obviously of great "humanistic" importance. For it means that the problem of the nature of reality is not prejudged. We are no longer required to believe that our response to beauty, or the mystic's sense of communion with God, have no objective counterpart. It is perfectly possible that they are, what they have so often been taken to be, clues to the nature of reality. Thus our various experiences are put on a more equal footing, as it were. Our religious aspirations, our perceptions of beauty, may not be the essentially illusory phenomena they were supposed to be. In this new scientific universe even mystics have a right to exist.

The outlook just described may fairly be said to be a result of the new scientific self-consciousness. It is more than a mere speculation. But some of the other views that have been put forth by our scientific philosophers seem a good deal less secure. We have, for instance, Sir James Jeans' view that the universe is a thought in the mind of a Supreme Mathematician. His reason for thinking this seems to be that *"all* the pictures which science now draws of nature, and which alone seem capable of according with observational fact, are *mathematical* pictures." [1] Moreover, these mathematical pictures are not pictures of anything that we can imagine. On the wave theory of matter, for example, an electron is a system of waves in a three-dimensional space. This sounds intelligible. We can identify this space with the physical space of our perceptions, and imagine the waves as being waves in some kind of ether. But we find that two electrons require a six-dimensional space, three electrons nine dimensions, and so on. It is evident that the space being talked of can have nothing to do with the space of perception. Also, we find that the waves being talked of are extremely elusive. It is suggested that they may be waves of probability, with no material existence whatever. It is obvious that the wave theory of matter is a description of something which is utterly unimaginable. From this, and similar instances, Sir James Jeans concludes that the universe is more like a thought than it is like anything else. He concludes, indeed, that it is a thought in the mind of a supreme mathematical thinker.

One objects to this argument, in the first place, that the universe possesses aspects other than its mathematical aspects. The artist and the mystic are concerned with aspects of the world which are not mathe-

[1] *The Mysterious Universe.*

matical. Even the sciences, excepting physics and its allies, are dealing with phenomena which are certainly not obviously mathematical. Adopting Jeans' line of argument, we should have to say that the Creator is thinking of a great many things besides pure mathematics. It might be said, of course, that all aspects of nature except its mathematical aspects are contributed wholly by our minds, and so are entirely subjective. This was very much the outlook of the early scientific creators, Galileo and Kepler.

But, if we adopt this way of reasoning, it is not at all obvious that nature's mathematical characteristics are not also subjective. Mathematical characteristics, it may be argued, are put into nature by us. We inevitably arrange phenomena in a mathematical framework because of the structure of our minds. This was Kant's view, and it is also Eddington's view when he says: "We have found that where science has progressed the farthest, the mind has but regained from nature that which the mind has put into nature." Jeans also thinks that our minds think mathematically from their very construction, but he regards it as a significant coincidence that nature should also behave mathematically, and finds in that coincidence evidence that nature has a mathematical Designer. But if Kant's views be correct, there is no coincidence, and the fact that we arrange nature in a mathematical framework tells us nothing at all about its designer.

Another point of view, which also has a respectable

philosophic pedigree, maintains that mathematics is not *a priori,* is not an inescapable activity of the mind. These philosophers maintain that the laws of mathematics are derived from experience. Mathematical thinking, they assert, has resulted from our observation of the actual laws obeyed by phenomena. Thus the fact that we think mathematically, and that nature works mathematically, is not an extraordinary coincidence, but simply an example of adaptation. Indeed, Mr. Bertrand Russell tells us that it can be shown that a mathematical web of some kind can be woven about any universe containing several objects. If this be so, then the fact that our universe lends itself to mathematical treatment is not a fact of any great philosophic significance.

But even if we agree that the universe has a mathematical designer, the next step, that the universe is a thought in a mathematical mind, seems a very long one. Sir James Jeans appears to have been led to this conclusion by the difficulty of imagining anything material behaving in accordance with the equations that modern physicists have found. He says:

> The concepts which now prove to be fundamental to our understanding of nature—a space which is finite; a space which is empty, so that one point differs from another solely in the properties of the space itself; four-dimensional, seven and more dimensional spaces; a space which for ever expands; a sequence of events which follows the laws of prob-

> ability instead of the laws of causation—or, alternatively, a sequence of events which can only be fully and consistently described by going outside space and time, all these concepts seem to my mind to be structures of pure thought, incapable of realization in any sense which would properly be described as material.

The state of affairs described by Jeans certainly makes it likely that whatever it is that behaves in this extraordinary way is not something that we can represent to ourselves in terms of familiar concepts. The "atom" of Victorian science could be pictured as a tiny grain of sand. The "electron" of modern physics certainly cannot be pictured as an even tinier grain of sand. But surely this is not surprising? Why should not the intimate workings of nature outrun our capacity for pictorial representation? Lord Kelvin said that he could understand nothing of which he could not make a mechanical model, and for that reason he never accepted Maxwell's electromagnetic theory of light. That seems to us now a strange criterion. Why should a man suppose that nature must be the kind of thing that a nineteenth-century engineer can reproduce in his workshop? What degree of unfamiliarity will Sir James Jeans permit before he declares that nothing material could act so oddly, and that the universe must be pure thought? It is also possible, we must remember, that some of these amazing mathematical webs that have been woven testify to the inefficiency of the mathematicians. It is perfectly possible that some of their paradoxes arise from the fact that they are using inappropriate concepts, and that the subject will take on a much more coherent form when new concepts are introduced.

Nevertheless, although they adopt very different routes, both Eddington and Jeans arrive at very much the same conclusion, namely, that the ultimate nature of the universe is mental. We have seen that Jeans has been led to this conclusion by the impossibility of conceiving anything save pure thought to which the modern mathematical description of the universe could apply. Eddington reaches his conclusion by reflecting that the only direct knowledge we possess is knowledge of mental states. All other knowledge, such as our knowledge of the material universe, is inferred knowledge—often the product of a long and complicated chain of inference. He holds the well-known theory that all our knowledge of the external world comes to us in the form of physical stimuli which travel along the nerves to the brain. Having arrived at the brain these stimuli are somehow transformed into, or give rise to, mental states, which are apparently of an entirely different nature from the physical stimuli. But, he argues, the only link in this chain of whose *nature* we know anything is the last link—the mental state. Are we to suppose that this link is something absolutely different in kind from the other links?

Consider vision, for example. The process starts

with the vibrating atoms of an external object. These vibrating atoms give rise to what the older physics called waves in the ether, but which the newer physics cannot yet satisfactorily describe. This physical process, whatever it may be, reaches the eye, and causes another physical process there. This, in turn, leads presumably to molecular movements in the brain. There then occurs the whole dissimilar phenomenon we call seeing a glowing red patch. This phenomenon appears so unlike those that preceded it that an absolute breach of continuity seems to have occurred. But why should we suppose this? We have seen that the only knowledge we have of the vibrating atoms, ether waves, and so on, is knowledge of their *structure*. This tells us nothing about their nature. May it not be, then, that their nature is the same as that of the red patch, namely, mental? This is, in essence, what Professor Eddington asserts. He quotes with approval W. K. Clifford's remark: "The succession of feelings which constitutes a man's consciousness is the reality which produces in our minds the perception of the motions of his brain." Seen from the outside, as it were, a living brain is a collection of molecules in movement. Experienced from the inside, it is a collection of mental states. The first view gives us knowledge of structure. The second view gives us knowledge of nature or substance.

Thus for Eddington the whole of what exists, the external universe and our minds, is homogeneous in its nature. "The stuff of the world," he says, "is mind-stuff." He goes on: "The mind-stuff is not spread in space and time; these are part of the cyclic scheme ultimately derived out of it. But we must presume that in some other way or aspect it can be differentiated into parts. Only here and there does it rise to the level of consciousness, but from such islands proceeds all knowledge. Besides the direct knowledge contained in each self-knowing unit, there is inferential knowledge. The latter includes our knowledge of the physical world." It will be seen that Eddington leaves his notion of "mind-stuff" very indefinite. It is not "stuff" since it is not spread in space and time, and it is not mind in the ordinary sense since it is only here and there that it rises to the level of consciousness. It may be regarded, perhaps, as an extension of the "unconscious" that psychologists talk about. Mr. C. D. Broad's very able analysis of the "Unconscious" in his *The Mind and Its Place in Nature* shows us the lines on which this concept could be developed. Jeans, as we have seen, goes even further, and would make the universe even more fully mental, being, indeed, a thought in the mind of God.

The humanistic importance of this outlook, in the minds of its authors, seems to be that it leaves us more free to attach the traditional significance to our æsthetic, religious or, compendiously, mystic experiences. It does not actively reinforce any particular religious interpretation of the universe, but it cuts the ground from under those arguments which were held to prove that any such interpretation is necessarily illusory. This it does by showing that science deals with but a partial aspect of reality, and that there is no faintest reason for supposing that everything science ignores is less real that what it accepts. The question as to why science can afford to ignore these other elements has also been answered by Eddington. Why is it that science forms a closed system? Why is it that the elements of reality it ignores never come in to disturb it? The reason is that all the terms of physics are defined in terms of one another. The abstractions with which physics begins are all it ever has to do with. By starting with "point-events," for instance, we can, mathematically, grind out one expression after another until we come to the mathematical specification of "matter." From this specification we can continue until we arrive back at our starting point.

We are doing what the dictionary compiler did when he defined a violin as a small violoncello, and a violoncello as a large violin. But what we have left out of this description is the process by which the mind of the scientist makes contact with one of the entities, namely, matter, which appears in this mathematical chain. It is in virtue of his recognition of this entity that the cycle of definitions has a meaning for him, and gives him genuine information. Similarly, to one who has seen either a violin or a violoncello, the dictionary definition gives information. But as long as the abstractions of physics form a closed cycle it is obviously immune from all disturbance from factors it has neglected.

It is not quite clear that this immunity will endure. The above analysis applies only to "field" physics which, as we have mentioned, covers a very large part of physics. But it does not cover the whole of physics, and the hope that it could be made to do so grows steadily less. In atomic and subatomic phenomena we seem to be faced by a state of affairs that lies right outside the cyclic scheme. The most disconcerting characteristic of this region is that strict causality, a cardinal assumption in science, does not seem to apply. In the motions of individual atoms and electrons there seems to be an element of free-will. Determinism has broken down, and the principle of indeterminacy has taken its place. There is great difference of opinion at present as to whether this is a genuine discovery, or as to whether it is a merely temporary technical device. Einstein, Max Planck, and others, think that strict causality will ultimately be restored in physics, while such men as Eddington and Schrödinger think that determinism must be definitely abandoned.

If the principle of indeterminacy comes to be definitely established, it will obviously have important philosophic consequences. It will make it easier to believe that our intuition of free-will is not an illusion. Moreover, instead of regarding the course of nature as the mere unrolling of a vast machine where every product is predetermined, we shall be freer to attribute to nature a genuine creative advance. And the distinction between the natural and the supernatural, as Eddington has pointed out, would be appreciably diminished. Indeed, if this principle be definitely admitted it will lead to the greatest revolution in scientific thought, and in the philosophy based on it, that has yet occurred.

So far we have dealt with the limitations of science as a method of acquiring knowledge about reality. We have seen that the new self-consciousness of science has resulted in the recognition that its claims were greatly exaggerated. The philosophy based on science had made "matter and motion" the sole reality. In doing so it had dismissed other elements of our experience, those that seemed to us to have the greatest significance and which, finally, made life worth living, as illusory. Science, in spite of all its practical benefits, had seemed to many thoughtful men, perhaps to the majority, to have darkened life. That the new attitude of science, as explained by such men as Eddington and Jeans, has obtained such widespread attention is not, therefore, surprising. It was the metaphysical doctrines that accompanied science that were found so depressing.

This is a striking testimonial to the fact that man is, after all, a philosophizing animal, for the purely practical aspects of science met with almost universal acclamation. And the wonderful vistas opened up by science seem to have been sufficient, in themselves, to inspire many men with optimism and a sense of freedom. Professor Millikan, an eminent man of science, has eloquently expressed this outlook. He points out that our civilization is the first one in history that has not been based on slave labour. This is unquestionably a great gain. Also the amenities of life are now more numerous and more widely distributed than they have ever been before. And we may have every confidence that this advancement will continue. The sciences of geology, palæontology, and biology have revealed an orderly development of living forms from the lower to the higher which has been going on for many millions of years. We may well suppose that it is part of the order of nature that such development should continue. Millikan quotes:

> A firemist and a planet,
> A crystal and a cell.
> A jelly-fish and a saurian
> And caves where the cave-men dwell;
> Then a sense of law and beauty,
> And a face turned from the clod—
> Some call it Evolution,
> And others call it God,

and comments: "That sort of sentiment is the gift of modern science to the world." And the fact that we ourselves are playing a part, a conscious part, in this evolutionary process, is, he thinks, of "tremendous inspirational appeal."

But it may justly be objected that all this is consistent with the old scientific philosophy that made man a purely accidental outcome of matter and motion, and which presented the universe, large and magnificent as it is, as entirely purposeless. The doctrine of progress, seen against that background, loses a good deal of its inspirational appeal unless we are content to find the whole meaning of life in the rearing of better and better human beings, until, with the final death of the physical universe, the whole irrelevant human adventure comes to an end. The matter is not much better if we suppose the physical universe to be continually renewed. Our religious impulses cannot be satisfied with anything less than a belief that life has a transcendental significance. And it is precisely this belief that the old philosophy of science made impossible. We conclude, therefore, that the truly significant change in modern science is not to be found in its increased powers to aid man's progress, but in the change in its metaphysical foundations.

CHAPTER VII

The Values of Science

SCIENCE is one of those human activities which have undergone a change of purpose or, at least, have come to serve different purposes, in the course of their development. The full wealth of values served by the major human activities are not apparent from the beginning. The cave drawings of primitive man, for example, are justly looked upon as the beginnings of pictorial art in the world. But it is very unlikely that primitive man conceived and executed them for purely artistic motives. They had, it is probable, a religious or magic significance. These representations of animals pierced with arrows were probably designed to secure a good kill on the morrow. And the early essays in science, as we have seen, were undertaken for purely practical reasons. Astronomy originated in the desire to construct a calendar for the benefit of an agricultural community, and the official reason given today for its support by civilized governments is that it aids navigation at sea.

Astronomy later developed into astrology, which had as its purpose the extremely practical one of foretelling human destiny. Biology began as medicine. Chemistry was for a long time alchemy, which was pursued chiefly because it was thought to hold the key to the discovery of the philosophers' stone and the elixir of life. Even mathematics was not always pursued for "its own sake." Certain mathematical relations were sought out for their supposed magical potencies, and others were valued for their usefulness in interpreting cabbalistic writings.

We have seen, however, that the sciences gradually lost their purely practical orientation and came to be pursued for the part they played in meeting man's curiosity about the external world. The "search for truth" became the dominant motive in the prosecution of science. In the development of science it was often found, of course, that answers to questions of purely theoretic interest were also of great practical importance. But these practical applications were, very often, mere by-products of the scientific movement. Thus Maxwell's discovery of the electromagnetic theory of light was not prompted by the desire to provide every home with a wireless set.

Disinterested curiosity has been the great motive power of scientific research. Of the great "values" that condition our activities and make our lives worth living, Goodness, Beauty, Truth, science has been chiefly concerned with truth. But "truth" does not seem to be a simple and unambiguous concept. We hear of "higher truths" and "deeper truths." We may ask then, What sort of truth is science after? In what sense is a true scientific statement true?

This question, which the characteristically modern scientific researches have raised to a position of supreme importance, has not yet received one clear, unequivocal answer. It will help us to understand the present position if we begin by considering what is meant by mathematical truth.

In the beginning, when man first became conscious of this curious mental power called mathematical reasoning, he regarded mathematical theorems as revelations of the fundamental principles on which the universe is constructed. And not only the exterior material universe, but also the universes of thought and emotion. Everything was regarded as being, essentially, an embodiment of number. Thus reason, whose great characteristic, according to the Pythagoreans, is its unchangeability, is a sort of embodiment of the number one. Mere opinion, as contrasted with reason, is an embodiment of the number two. The number two is also feminine, while three is masculine. Thus the essence of marriage is represented by the number five, which is a combination of the first feminine and the first masculine number. In this extreme form these ideas did not long survive. But, modified and debased, they persisted for centuries. We find traces of this outlook even in Kepler, and it persists to our own day, attenuated and incoherent, in superstitions about "unlucky numbers."

But even though Pythagorean extravagances had been abandoned, mathematics, until quite recent times, maintained the exalted position of acquainting us with "necessary truths." Descartes expresses this point of view in a famous passage from his Fifth Meditation:

> I imagine a triangle, although perhaps such a figure does not exist and never has existed anywhere in the world outside my thought. Nevertheless, this figure has a certain nature or form or determinate essence which is immutable and eternal, and which I have not invented, and which depends in no way on my mind. This is evident from the fact that I am able to demonstrate various properties of this triangle, for example, that its three interior angles are equal to two right angles, that the greatest angle is subtended by the greatest side, and so on. Whether I want to or not, I recognize very clearly and evidently that these properties are in the triangle although I have never thought about them before, and even if this is the first time I have imagined a triangle. Nevertheless, no one can say that I have invented or imagined them.

Thus the properties of a triangle, according to Descartes, in no way depend on the human mind. But the invention of the non-Euclidean geometries has taught us that these properties are necessary merely in the sense of being logical consequences of the axioms and postulates with which we start. And these axioms and postulates are arbitrary. They are not necessities of thought; they are matters of the mathematician's caprice. We can, if we like, start with the axioms assumed by Euclid, and then we get the properties that so impressed Descartes. Or we can start with Lobachevsky's axioms, and then we get quite different properties. Or we can start with Riemann's axioms, and get still different properties. And so on. None of these axioms are necessities of thought. Regarded logically, there is nothing to choose between them. Thus there is no reason to suppose that Descartes's triangle is a revelation of an eternally pre-existing truth—such as a thought in the mind of God. It is an arbitrary creation of the mathematician's mind, and did not exist until the mathematician thought of it.

Although the idea has been abandoned that mathematics gives us access to some external store of supersensible truths, there is still controversy as to its real nature. A school of continental mathematicians holds that it is a sort of game, like noughts and crosses. Another school holds that it is a branch of formal logic. The arguments about this matter are extremely subtle and fatiguing, and the melancholy fact that they are not also convincing is sufficient testimony to the difficulty of the subject.

But the negative result of these investigations is sufficient for our present purpose; they have established the fact that mathematics has not the transcendental significance that was attributed to it by philosophers for about two thousand years. At the same time the precise nature of mathematical entities is still a difficult and controversial subject. Although the simple, primary mathematical ideas were doubtless originally suggested by experience, the mathematician's development of them has been very largely independent of the teachings of experience. He has been guided chiefly by considerations of form—a criterion which is probably, at bottom, æsthetic.

This development has, in fact, been so largely self-contained, so autonomous, that the fact that mathematical theorems can be applied to the happenings of the real world seems like a lucky accident. This lucky accident is so remarkable that it has led some modern speculators, as we have seen, to conclude that the universe had a mathematical designer. As against this we must remember the possibility that a mathematical web of some sort could be woven about any universe, and the extraordinary difficulties and incoherencies of modern mathematical physics suggests that, in spite of the extraordinary luxuriance of mathematical creation during the last few centuries, the universe is so odd an affair that we are not yet in a position to weave a satisfactory mathematical web about it. To sum up, we may say that, even if mathematics be not purely subjective, it is not, in its nature, a body of truth about the objective, physical world. If it has an objective reference it is only in the sense that logic, according to some writers, has an objective reference.

When we come to science proper, as in the science of physics, we are obviously in a region where experience has played a much greater role. We have seen that all its primary concepts have been suggested by experience. In fact, to say that they have been "suggested" by experience would seem to many an understatement. In their original intention, at any rate, the fundamental terms of physics were put forward as names for certain actual characteristics of nature. The physicist gave names to certain qualities that he observed just as Adam gave names to the animals in the Garden of Eden.

When Newton isolated the concept of "mass," for example, he thought he was singling out and naming an actual objective constituent of the external universe. Other properties of matter, such as size, shape, position, velocity, were also regarded as obviously objective existents. Force was a rather more mysterious notion. It was derived from our sensation of muscular effort. But it was not at all clear how we were to regard two distant bodies as pushing and pulling one another. It was hoped, however, to reduce all forces to impacts. It was hoped that, in the last analysis, all forces would reduce to the effects of little hard particles knocking against one another. Another way of dealing with the mystery was to declare the notion of force superfluous, and attempts were made to construct systems of mechanics which would not employ this notion. Thus force came to be regarded as a semi-fictitious entity; it was a more abstract concept than the straightforward notions of mass, velocity, and so on. It became doubtful whether all these physical concepts are just names for objectively existing entities.

This doubt is heightened when we come to consider some of the entities that were later imported into physics. Thus the notion of energy is one of the most important of all physical concepts, and the principle of the Conservation of Energy one of its most important principles. But what is the precise status of this principle? Is it a fact of observation that there is some constituent of the external universe to which we have given the name Energy, and is it a fact of observation that this characteristic never increases and never diminishes?

Let us consider, for example, what happens when we let a stone fall from a height. The stone, on colliding with the ground, acquires heat, and the earth for a certain distance round the stone also acquires heat. Thus the universe, it appears, possesses more energy than it had at the moment before we let the stone fall. Where has this energy come from? It has come, we are told, from the energy of motion of the stone. Every moving body possesses energy in virtue of its motion. The stone has steadily been acquiring this sort of energy during its fall, and at the moment of impact, when its energy of motion is at its maximum, this energy is changed into heat energy.

But from what source has the stone been steadily acquiring its energy of motion? Here we come to the somewhat mysterious notion of "potential" energy. A stone, or any other body, we are told, can possess energy purely in virtue of its position with respect to surrounding bodies. Thus a stone at the top of a mountain possesses more energy than an exactly similar stone at the foot of the mountain. Exactly similar stationary stones at different levels possess different energies. These energies are the potential energies of the stones. Potential energy, it must be admitted, is a somewhat mysterious notion. Other forms of energy, such as energy of motion and heat energy, are obviously "energetic." But potential energy is undetectable until it is transformed.

But it is only by importing this notion of potential energy that we can assert the conservation of energy. Otherwise, as in the case of the falling stone, we seem able to create energy. Similarly, by throwing a stone up in the air, we seem able to annihilate energy. At the moment of leaving the hand, the stone possesses energy of motion given to it by muscular exertion which has consumed energy from the body. But when the stone is at the top of its flight, its energy of motion, which has steadily been becoming less, becomes zero. It then begins again to acquire energy of motion, until it reaches the hand with the same energy with which it left it. The notion of potential energy comes in here to preserve the principle of the conservation of energy by saying that, as the stone's energy of motion decreases in its upward flight, its potential energy increases in such a way that, at the top of its flight, the potential energy is equal to the energy of motion with which the stone started.

Thus the notion of potential energy explains away

apparent violations of the principle of the conservation of energy. But is not this the very reason for the importation of the notion of potential energy? Is it not a mathematical fiction, brought in for convenience? A quotation from a well-known Victorian text-book of physics[1] will show us the sort of attempts that were made to justify potential energy as a real physical quantity.

> The question still remains, what becomes of the motion when the kinetic energy of a system diminishes? Can motion ever be changed into anything else than motion? If we assume a fundamental medium whereby to explain all the phenomena of nature, then the properties of this medium ought to remain unchanged, and all other changes must be explained by motion of the medium. Such an assumption is quite philosophic, and the method of procedure is certainly scientific. An evident reply to the question of what becomes of the motion of a projectile rising upwards is that it passes into the ether. The first assumed property of the ether is that it can contain and convey energy. There is no *a priori* reason, then, why the energy of motion of a projectile as it rises upwards should not be stored up as energy of motion of the ether between the body and the earth, or elsewhere. The oscillation from kinetic to potential, and from potential to kinetic, in the case of the pendulum is then, from this point of view, merely an interchange of energy of motion going on between the mass of the pendulum and the

[1] Preston's *Theory of Heat.*

ether around it. According to this view, all energy is energy of motion, and must be measured by the ordinary mechanical standard. The work we do in lifting a body from the earth is spent in generating motion in the ether, and as the body falls this motion passes from the ether to the body, which thus acquires velocity.

A rough mental picture of the process might be obtained as follows: We might suppose a body connected to the earth by vortex filaments in the ether, which would replace the lines of force. The ether is spinning round these lines, and when the body is lifted from the earth the work done is expended in increasing the length of the vortex filaments. The work is thus being stored up as energy of motion of the ether, and when the body falls to earth the vortex lines diminish in length, and their energy of motion passes into the body and is represented by the kinetic energy of the mass.

This quotation is sufficient to show that the author is uneasy at the apparently conventional character of potential energy, and accordingly invokes unknown properties of the ether in order to give it standing as a physical quantity. This line of reasoning raises the principle of the conservation of energy to a position where it can no longer be either verified or disproved. Thus the author in question goes on to say that, if a material system is found to be losing energy in some way that we cannot account for, then we must assume that the system is radiating energy into the ether in some unknown manner. Similarly, if the system is found to be gaining energy, we must assume that the ether is somehow conveying energy to it.

Thus the conservation of energy becomes an article of faith. It is no longer an observed characteristic of nature, based upon experimental evidence, but becomes, at best, a metaphysical doctrine. We see how difficult it is, in practice, for science to remain faithful to Newton's criterion. In modern relativity theory the notion of potential energy is not used, and the principle of the conservation of energy is somewhat modified. We find that the notion of potential energy was not an arbitrary creation; it arose from a sound mathematical intuition, but the problem it was invented to answer was not correctly formulated. The art of putting correct questions to nature is learned only gradually, and there can be no doubt that some of our present difficulties arise from the fact that this art is not yet completely mastered.

§ 2

The examples of force and potential energy suffice to show that the naive view of science is not correct, and that science does not construct its mathematical formulation of the world wholly in terms of directly observable physical quantities. There is, in science, a certain amount of useful myth. But the myths are useful because they are, as it were, pegs on which the mathematical formulation can be hung. It is very seldom, however, that myths have been introduced in full consciousness of their mythical character. Thus it can hardly be doubted that Newton believed in the reality of forces. Potential energy, as our quotation shows, was regarded as more than a mathematical device. Bohr's theory of the atom, with its circulating electrons, was at first taken to be a faithful picture of the objective reality. It was only later that it came to be realized that any other picture will do equally well provided it leads to the same mathematical equations. And the modern Schrödinger atom, with its waves in multi-dimensional space, presents us with a picture which is frankly incredible. Again all that matters is the mathematical formulation.

The reason for this apparently curious unconsciousness of the true nature of their task, on the part of men of science, is very simple and obvious. It is that men try to explain the unfamiliar in terms of the familiar. Scientific concepts were suggested by ordinary daily experience. It was only gradually discovered that natural processes cannot be accurately described in terms of these concepts. Whether man will ever be able to evolve concepts which are entirely adequate to nature is obviously a question which cannot yet be decided. The materials out of which he has built his theories have hitherto shown themselves to be not quite suitable. It is as if one were given the task of constructing a map of a country out of little straight sticks. It was impossible, for instance, for Newton to give a perfectly accurate picture of the solar system with his apparatus of forces and Euclidean geometry, just as the Ptolemaic apparatus of circular motions *could* not be made to fit the observed heavenly motions. Similarly, the motion of a little hard particle, however much it may be refined, cannot be made to account for the behaviour of an electron. Even such guiding principles of thought as continuity and causality are being found inappropriate for the scientific description of nature.

It is evident, even from this brief survey of scientific ideas, that a true scientific theory merely means a successful working hypothesis. It is highly probable that all scientific theories are wrong. Those that we accept are verifiable within our present limits of observation. Truth, then, in science, is a pragmatic affair. A good scientific theory accounts for known facts and enables us to predict new ones which are then verified by observation. Thus Newton's theory of gravitation accounted, within the then limits of observation, for the heavenly motions. Subsequently, in the hands of Adams and Leverrier, it enabled the existence of a hitherto unknown planet, Neptune, to be predicted. It did, in fact, lead a very successful career, and is as good an example of a scientific theory as can be found. But sufficiently precise observation showed that it did not perfectly account for the motion of the planet Mercury, and it was then seen that its account of the motions of the other planets only seemed complete because, in those cases, departures from the predicted motions were below the limits of observation.

The way in which scientific men attempted to remedy the defects in Newton's theory is instructive. Their method was to tinker with the details. They tried the effect of making the law of attraction just a little different from the exact inverse square law. They tried the effect of giving gravitation a finite velocity of propagation. But no advance was made in this way. The difficulties could not be met by any extension of the theory. It was not until Einstein introduced an entirely new theory, based on entirely new concepts, that the difficulties were cleared up. We now know that the apparatus of ideas with which Newton started, however deftly used, cannot be made to perform the task for which it was invented. It would probably be impossible to find another instance of a change as fundamental as this in the history of the sciences. Here we have no step by step modification of a theory, but the complete sweeping away of the very foundations of a scientific outlook. The one feature of the Newtonian outlook that is left is the insistence on a mathematical formulation of nature.

Another instance, although rather less marked, of a radical change of outlook, is provided by the fate of the ether theory. The universal medium called the ether was postulated chiefly in order to account for the propagation of light. Light takes eight minutes to reach us from the sun. How is it transmitted? The hypothesis that it consists of little corpuscles shot out of the sun was found to be unsatisfactory, and the theory was developed that light is a wave motion of some kind. But a wave motion implies, of course, a medium in which the waves travel. Thus there naturally arose the question as to the constitution of the medium. It was at first conceived, on the basis of previous experience of extended and not readily perceptible media, as a sort of gas. Certain experiments showed that this theory was inadmissible, and the notion of the ether as a sort of huge jelly was then developed. This fitted the observed facts much better—but the accordance was not perfect. Attempts to perfect the ether theory led to the amazing developments that we have already described.

Then, by some obscure process of reason and intuition that cannot be clearly analysed, Maxwell developed his equations. He started with mechanical concepts not unlike those of his predecessors, but in developing them he made jumps—flashes of genius—that took him outside the mechanical scheme. He had arrived at the correct mathematical formulation of light-processes. He had thus, in a sense, reached the goal at which all the ether theorists had been aiming. But his mathematical formulation was not reducible to mechanical terms. Then what is Maxwell's theory? Is it true, as Hertz said, that "Maxwell's theory is Maxwell's system of equations"? In view of recent controversies this question conceals an ambiguity which it is desirable to make plain.

The issue raised divides physicists into two camps, and has an important bearing on our question as to what is meant by truth in science. The one group holds what is called the "conventionalist" view, which has been expressed by a German writer as follows:

> It has sometimes been supposed that a physical magnitude, such as time, for example, may have a meaning in itself, without any reference to the way in which it is to be measured, and that the method of measurement is a second question. As opposed to this view, it must be strongly emphasized that the meaning of every physical magnitude consists in certain numbers being ascribed to certain physical objects. Until it has been settled how this ascription is to take place, the magnitude itself is not settled and statements concerning it have no meaning.

Thus, on this view, there is no such thing as a physical quantity apart from a system of measurement. The other group holds that there are objective physical quantities which exist quite independently of our methods of measuring them. We may call this the "realist" view. The conventionalist outlook is obviously an extension to physics of the outlook of the pure mathematician, for whom an entity exists purely in virtue of its definition. But the outlook seems to assign altogether too great a role to the subjective element in science. It may be true that the square root of 2 had no existence before the mathematician thought of it, but it certainly seems to be untrue that lapses of time, variations of light intensity, and so on, did not exist before the scientist devised ways of measuring them. We cannot say that the child or savage has no knowledge whatever of such things simply because he cannot measure them. It seems, rather, that there are objective physical quantities of which he has a confused apprehension compared with the more precise apprehension of the scientific man.

And in the history of scientific measurement itself, the change from one system of measurement to another certainly does not appear to be a purely conventional change, but appears to be undertaken in the effort to obtain a more precise estimate of the magnitude of the physical quantity concerned. Thus, according to the conventionalist, the temperature of a body must be the reading on a thermometer. But scientific men, in their search for accuracy, have successively used different thermometers. Are there, then, as many temperatures as there are thermometers? Is there no objective physical quantity, *the* temperature of the body, that scientific men are trying to reach? According to the conventionalist it is meaningless to talk of any such quantity.

Another objection to the conventionalist outlook comes from the fact that, in the investigation of a physical quantity, entirely different methods, based on entirely different conventions, reach the same result.

Unless it can be shown that the conventions have been specially designed so as to fit in with one another, we can explain the agreement only by supposing that there is an objective physical quantity that the measurements measure. But even if we assume that the world of physical quantities is a world that is investigated, but not created, by the physicist, it still remains true that the only knowledge of this world that science gives us is knowledge of its mathematical aspects.

The conventionalist view, we have given reason to suppose, goes too far. Nevertheless, there are physical quantities to which it can be plausibly applied. We have given instances in the notions of "force" and "potential energy." These, it is evident, are not constituents of the external world in the sense that mass, temperature, electric charge, are constituents of the external world. But the basis of this difference is not to be found in our sense impressions. The earliest quantities isolated by physics are precisely those that directly affected our senses. Even "heat" and "cold," as we have seen, were at one time regarded as distinct entities. But the degree of reality we now attach to a physical quantity seems to depend upon the part it plays in the coherent scheme of physics. And the history of physics shows that it is precisely in its search for reality that physics is led to its most abstract abstractions. And all that physics can tell us about these abstractions, we must repeat, is their mathematical specifications. Thus the "real" world of physics is vastly different from the world of perception.

Owing to the lack of connexion between quantum physics and field physics, it is impossible to say at present what are the fundamental physical quantities out of which physics proposes to build the world. Field physics, it is possible, can be based on "point-events" and the relation called the "interval." This will not enable us to deduce quantum phenomena, which are at present based on different fundamental entities. But it is evident that, if we make an assumption similar to that of the old materialists, and regard the "real" world of physics as the only objectively existing world, then it is not only the "secondary qualities" of the old theorists that we must regard as illusory. Their fundamental realities must go too, for even space, time, and matter are all derivative from more fundamental entities, according to the new physics.

The science of physics is concerned only with those aspects of the world which can be treated quantitively. This imposes an immense restriction on the elements we select from the total elements of our experience. If we go further, and assume that only the elements of which physics treats are real, all the others being in some way illusory, then we are led to some such apparently fantastic conclusion as that "point-events" are the only objective realities. But we may leave the philosophical issue involved here to the philosophers. Science itself does not assume that it alone is concerned with the real. The statements of science are true within the limited region it claims for itself—the region of mathematical structure. If you are curious about the mathematical structure of the material universe, then science can give you information on that point.

In pursuit of its aim science has shown itself to be accommodating and flexible. The science of physics has shown itself willing, at different times, to adopt quite different principles and concepts provided they promised to be helpful in its one great aim of giving a mathematical description of nature. Where principles and concepts have persisted past their time of usefulness, this has been due to the inertia of mental habit. No scientific principles are sacrosanct; no scientific theory is held with religious conviction. Nevertheless, most scientific men believe that there is a final scientific truth about the universe to which successive scientific theories ever more closely approximate. But this is an article of faith. Science is still an adventure, and all its "truths" are provisional.

§ 3

This atmosphere of provisional hypothesis and practically verifiable statements constitutes what has been called the "homely air" of science, and is one of its great charms. Science has adopted the pragmatic criterion of truth, namely, success, and as a result science has been successful. Indeed, it would not be too much to say that, judged by the criterion of general assent and practical efficacy, it is the most successful activity that man has yet hit upon. As Professor Levy [1] says:

> Any such criterion of truth may not be one that commends itself to the professional philosopher. It may be that science ignores subtleties that appear vital to academic philosophy, that it skims easily over the surface of reality. The very principle that scientists must leave no differences behind may narrow the range to superficial agreement, and restrict the nature and number of the isolates it may form. Whether or not this be so, science can at any rate look upon itself as a united movement that has left in its wake a body of tested knowledge, while philosophy is still broken up into disunited schools of thought.

The criticism has, indeed, been made that science pays for its success by its superficiality. All the deepest problems of mankind, it has been pointed out, lie outside science. If neither philosophy nor religion can present any such "body of tested knowledge," it is because they have not been content with such cheap victories. There is doubtless some truth in this criticism, and it is probably true that the problems with which science deals are intrinsically inferior in human interest to those dealt with by either philosophy or religion.

Nevertheless, the actual atmosphere of science, the

---

[1] *The Universe of Science.*

manner in which it goes about its work, is quite exceptionally agreeable. It is in the scientific attitude, as much as in the scientific results, that the true value of science is to be found. If the man of science has not aimed high, according to the philosopher, he has at least aimed with a single heart, with a docility in face of the facts, with an impersonal purpose to serve which is not always found amongst our prophets and philosophers, and which it is almost impossible to find elsewhere.

There is probably no other period in history, since modern science began, when the particular values it incorporates were so rarely to be encountered in other human activities. The human tendencies to prize certitude and fear knowledge, to indulge emotion at the expense of reason, were probably always as strong as they are today, but the circumstances of the time did not show them up in so pitiless a light. The mass of men were probably always as impatient of that careful, honest verification that is the very essence of science as they are today, but there was no huge popular press to bear daily witness to the fact.

If we are to judge from what seems the overwhelming evidence provided by such activities as politics, business, finance, we must conclude that the attention and respect accorded to science is directed wholly to its results, and that its spirit is the most unpopular thing in the modern world. Yet it could very reasonably be claimed that it is in its spirit that the chief value of science resides. This can be asserted without abating anything of our claim for the values of its results. Knowledge for the sake of knowledge, as the history of science proves, is an aim with an irresistible fascination for mankind, and which needs no defence. The mere fact that science does, to a great extent, gratify our intellectual curiosity, is a sufficient reason for its existence.

But it must be pointed out that this value is inextricably intertwined with another value—the æsthetic value of science. If scientific knowledge consisted of a mere inventory of facts, it might still be interesting and even useful, but it would not be one of the major activities of the mind. It would not be pursued with passion. It could, at best, only exert a somewhat more intensified form of the attraction we feel for a time-table. Indeed, the greatest single testimonial to the fact-gathering power of science, resting, as it does, on centuries of labour and ingenuity, is probably the *Nautical Almanac*—useful, but not entrancing, reading.

For science to have inspired such ardour and devotion in men it is obvious that it must meet one of the deepest needs of human nature. This need manifests itself as the desire for beauty. It is in its æsthetic aspect that the chief charm of science resides. This is true, be it noted, for scientific men themselves. To the majority of laymen, science is valuable chiefly for its practical applications. But to all the greatest men of science practical applications have emerged incidentally, as a sort of by-product. This is, perhaps, most obviously true of the men who have created the mathematical sciences. In the work of mathematicians, in particular, the æsthetic motive is very apparent. Many mathematicians have written about their work in a sort of prose poetry, and the satisfactions they get from it seem indistinguishable from those of an artist. The language of æsthetics is never far to seek in the writings of mathematicians. In their frequent references to the "elegance," "beauty," and so on of mathematical theorems they evidently imagine themselves to be appealing to sensibilities that all mathematicians share. Nearly all mathematicians show themselves uneasy in presence of a proof which is inelegant, however convincing—one which "merely commands assent," as Lord Rayleigh said—and, sooner, or later, endeavour to replace it by one which approaches closer to their æsthetic ideal. Some of them, as the late Henri Poincaré, have gone so far as to remark that the actual solution of a problem interests them much less than the beauty of the methods by which they found that solution.

There can be little doubt that it is the æsthetic element in mathematics that has been chiefly responsible for the attention it has received. If mathematics is to be ranked as a science, then it is, of all the sciences, the one most akin to the arts. The history of its development bears out this assertion, for this development, in the main, has been a formal development. Certain branches of mathematics, it is true, have been created to deal with some actual physical problem but, for the most part, mathematics has been a remarkably autonomous activity. In this respect it greatly resembles the art of music, whose development also has been so largely conditioned by considerations springing from its own ground, and owing nothing, or almost nothing, to the external world.

But if the æsthetic element is most prominent in mathematics, it is certainly not lacking in the physical sciences. The search for universal laws and comprehensive theories is indubitably a manifestation of the æsthetic impulse. It is sometimes said that scientific men, in their choice of one of two theories that equally well cover the facts, always choose the simpler. But simplicity is, at bottom, an æsthetic criterion. Fresnel said: "Nature takes no account of analytical difficulties." If we interpret this to mean that there is no *a priori* reason to suppose that nature is simple, then the remark is doubtless true. But if nature was not simple, it is very doubtful if science would ever have been pursued. It would still be possible for science, of a kind, to exist. It is within the resources of mathematics to construct a formula for almost any sequence of changes, however arbitrary, but if this formula is of a kind from which no further deductions can be drawn, the whole investigation is regarded as being in a very unsatisfactory state. Where such formulae are used, as they often are in engineering, for example, it is solely for their immediate practical value. The exclusive construction of such empirical formulae is not an object that any scientific man would propose for himself.

It is because nature does seem to exhibit an order we can understand, because diverse phenomena can be grouped under general laws, that the pursuit of science is worth while. If nature did not possess a harmony that was beautiful to contemplate, said Poincaré, science would not be worth pursuing, and life would not be worth living. Life would not be worth living, that is to say, for Poincaré. But although most scientific men would doubtless find a scientifically unæsthetic universe quite worth living in, we may be confident that they would not devote themselves to science. "You are young, I am old," wrote Faraday to Tyndall, "but then our subjects are so glorious, that to work at them rejoices and encourages the feeblest; delights and enchants the strongest." And Einstein, writing on Max Planck, reveals himself when he says:

> The longing to see this pre-established harmony is the source of the inexhaustible patience and persistence which we see in Planck's devotion to the most general problems of our science, undeflected by easier or more thankful tasks. I have often heard that colleagues sought to trace this characteristic to an extraordinary will-power; but I believe this to be wholly wrong. The emotional condition which renders possible such achievements is like that of the religious devotee or the lover; the daily striving is dictated by no principle or programme, but arises from an immediate personal need.

When we come to the sciences other than physics, we find the same spirit. Whether it be chemistry, biology, or what not, we find that all the great advances have come from the desire to uncover what Einstein calls "this pre-established harmony." Science can, of course, content many appetites. It enables many more millions of lives to exist than would be possible without it. Also it enables many more millions of lives to be destroyed than was possible without it. It can give us marvels to gape at: the distances of the stars, the minuteness of the atom. It can stun us with figures. It can even help the dividend-hunter in his absorbing quest. But to the great man of science, science is an art, and he himself is an artist. And his creation is not the less a work of art because it is but a faint and imperfect copy of another—of the supreme work of art which is nature itself.

§ 4

The æsthetic appeal of science is due to the fact that it bestows comprehension. Under the phrase, "æsthetic emotion," a number of elements are lumped together, and some only of them are present in any particular case. Our reaction to any great work of art is complex. But one of the most important of our reactions, in many cases, is that our desire for comprehension is gratified. To comprehend a thing is to see it in its relations, to see it in its place within a particular framework.

In the case of a work of literature this framework is

provided by the artist's personal vision of life. This is, as it were, the unspoken context within which everything that the artist says acquires a meaning. And our lasting reaction to a work of art, the degree to which it works in us and modifies us when we have forgotten all its details, is dependent on the depth and comprehensiveness of this vision. This is the light which pervades the whole work, and bestows on it such harmony as it possesses, a harmony which lies much deeper than anything the artist may achieve by the technical dovetailing of the elements of his work. The chief function of art is to communicate this vision, and it is the mystery and miracle of art that it can do so.

The comprehension bestowed by a work of art is really the communication of the artist's personal vision. The comprehension bestowed by science is not so obviously personal. Indeed, it is generally supposed to be wholly impersonal. The framework of reference, or "vision," within which a scientific theory exists is the group of fundamental concepts and principles in terms of which that region of nature is to be described. These fundamental terms have often been regarded as necessary and unchangeable. We have seen that they are not. Theoretically they have alternatives, although their adoption at that particular time may seem to have been psychologically inevitable.

The Newtonian concepts, for instance, were the final formulation of ideas which, in a more or less vague state, had been "in the air" for many years. Similarly, the biological theory of evolution has a long history. The Darwinian form of it, in fact, was independently hit upon by two different men. It was, as people say, "bound to come." Nevertheless, even if the impetus of the whole scientific movement was towards the Newtonian concepts, we see now that those concepts were not theoretically necessary. Einstein's utterly dissimilar way of looking at the same group of phenomena was a theoretically possible alternative. Not that Einstein could have produced his theory if he had lived at Newton's time. In the absence of the necessary mathematical technique that would have been impossible. In Newton's time it is probable that Einstein would either have developed Newton's theory or else have been the impotent prey of intuitions that he could not formulate. But this would have been merely an accident of history. Theoretically, the Newtonian concepts had alternatives. They were chosen although, as a mere matter of psychology, it may be, there was in fact no choice.

Einstein's theory is perhaps the clearest instance that can be given of our assertion that even a scientific theory may possess a personal element. Einstein's theory is so original that it is very difficult, even after the event, to provide it with an ancestry. It was not in the least a natural culmination of the ideas that preceded it. It was a bolt from the blue. The extraordinary lack of comprehension with which the scientific world at first greeted it was due not to its technical difficulties, but to the unfamiliarity of the outlook it assumed. It seemed to be the product of an alien mind.

We could say of this theory what Einstein said of some of the work of Gauss, that, if its author had not thought of it, there is no reason to suppose that it would ever have been thought of. Science provides few instances of theories as original as this—indeed, perhaps there is no other scientific theory so intensely original as Einstein's theory. But there are other scientific theories which have a strong element of originality. Maxwell's electromagnetic theory, for example, was a very lonely achievement.

And it is precisely this quality of originality which is the personal element in science. The majority of men of science, including some of the greatest, impress us as journeying along the highroad. Newton, for instance, revealed nothing totally unsuspected by the best of his contemporaries. It is as if many converging lines of thought came to a focus in him. Newton's great quality was not a bewildering originality, but a bewildering and unequalled mastery of his material. Other men could see his problems; only Newton could solve them.

But the men who strike us as having branched away from the highroad, and having thereby given science an entirely new direction, have put to themselves entirely new problems. It is here, of course, that the personal element in the creation of a scientific theory is most clearly revealed. The history of science is not the history of some sort of automatic development. The actual course that science has pursued depends very largely on the types of mind which, as historical accidents, happen to have risen to the level of genius at favourable instants.

It is often said that the essential distinction between science and art consists in the fact that science makes appeal to universal assent, whereas art does not. A scientific statement, we are told, is open to verification by anybody, whereas a work of art appeals only to people with certain sensibilities. A work of music means nothing to a man who is tone-deaf. Science deals with a "public" world, whereas art is concerned with a private world. A colour-blind man, for instance, would not appreciate painting, whereas a man born blind could master the whole theory of optics.

What are the primary judgments to which, according to this theory, science appeals? So far as the science of physics is concerned, they are judgments about the indications of measuring instruments. We have to say, for example, that a pointer coincides with a certain mark on a scale. We have to agree, also, about number judgments. We have to agree that an urn contains twelve balls, and not eleven or thirteen. These are the only judgments, we are told, that are involved in science, and about these we can secure unanimity. Two observers, with different colour perceptions, might disagree as to the colour of some particular bright line in the spectrum, but they would both agree as to its position on a numbered scale, that is, they would agree about its wave-length. And it is its wave-length, not its colour, that is dealt with by science. In any scientific statement where a colour is mentioned, the name of the colour could be deleted, and a figure for a wave-length substituted, without harming anything in the statement. The same is true about any scientific statement which seems to appeal to judgments about which men might differ. Two men might disagree as to whether one note was the octave of another, judging by sound only, but they would agree as to whether its rate of vibration, measured on a scale, was or was not double that of the other. And that, we are told, is all that is necessary for the purposes of science.

Now we may admit that universal agreement may be obtained about such things as the number of objects in a collection or coincidences in space without therefore concluding that science is potentially capable of securing universal assent. For science consists of a great deal more than such elementary judgments. The chief thing about science is its theories, and it is surely obvious that not all men are capable of assenting to its theories. A man may agree that a star crosses a wire in a transit telescope at the moment when the hands of a clock mark a certain time, but the theorem which enabled the astronomer to predict that occurrence may be for ever inaccessible to him. As a matter of fact, the upholders of this theory, which professes to be so realistic in the sense of dispensing with all "subjective" elements, are not realistic at all.

A man may be blind, deaf, dumb, and paralysed—that, they cheerfully maintain, does not matter at all. But he must be a very fine reasoner. If we adopt this criterion that the truth of science is to be decided by a unanimous vote, then we must point out this unanimity, when it comes to actual science, and is not confined to such questions as the number of matches in a match-box, cannot be obtained. A large section of mankind, perhaps the majority, are congenitally incapable of understanding Einstein's theory, for example. If such people are to be ruled out as not affecting the unanimous vote, on what ground is the poet to be differentiated who rules out all those people who are insensitive to poetry? In both cases the claim could be made that *if* all people possessed the requisite sensibilities or faculties, the vote would be unanimous. And this is, in fact, the implied claim of those who preach the complete objectivity of science. For their criterion for objectivity is merely universal assent.

But even if we abandon the criterion of universal assent, and grant votes only to those capable of forming a judgment, we still do not get unanimity. It is notorious that theories that have been found convincing by some scientific men have been found unconvincing by others. Faraday, for instance, was opposed to the atomic theory at a time when, in the judgment of most of his contemporaries, it was well established.

At the present day we have an interesting example of the influence of purely "subjective" factors in the creation of scientific theories in the methods adopted by Einstein and Eddington respectively in their attempts to reduce the laws of electromagnetism to geometry. Judged by the scientific criterion of accounting for phenomena there seems nothing to choose between them. But Einstein has said that he dislikes Eddington's theory, although he is unable to disprove it, and Eddington has said of Einstein's theory that it is a matter of taste. Here we are in a region where the ordinary "objective" criteria fail us. Our attitude towards these theories seems to depend on considerations which are, at bottom, æsthetic.

Other instances could be given, particularly from present-day quantum theory, where a chain of reasoning which is found quite convincing by one authority is found unconvincing by another. And the different judgments seem to depend on considerations which the authorities concerned find too elusive for expression. This dilemma sometimes becomes so acute, Professor Levy tells us, that a scientific society, in view of the conflicting reports of its referees, cannot decide whether to publish a certain paper or not. And Poincaré has told us that even in pure mathematics, where reason, one would think, is most pure and undefiled, a proof which is quite satisfactory to one mathematician is often not at all satisfactory to another. Indeed, Poincaré was led to divide mathematicians into psychological types, and to point out that a kind of reasoning which would convince one type would never convince another. These differences, in his opinion, are fundamental, and play a great part in the actual construction of science.

In view of these facts it is obviously misleading to present science as differing fundamentally from the arts by its "impersonal" character. There is no absolute difference here, but only a difference of degree. Science is less personal than the arts, but it is a mistake to suppose that it is wholly impersonal. The reason for the difference lies in the fact that beauty and truth, in spite of Keats, are commonly distinguished. A work of art aims more consciously and deliberately at beauty than does a scientific theory. There are some works of art, indeed, where it is difficult to see that the criterion of truthfulness has any relevance at all. This is most obviously the case with some musical works. There are other works of art—e.g. in literature and painting—where truthfulness is relevant. And while it is probably true that no scientific theory has been constructed in complete independence of æsthetic considerations, it is nevertheless true that these alone are not sufficient.

Perhaps this is as far as the discussion can be carried on the common-sense level. A further analysis would have to distinguish between, for example, the "truth" of mathematics and the "truth" of physics. And we would have to question the complete independence of the quality called "beauty." We have already likened mathematics to music, and it could be objected that, since a mathematical development must be logical, any æsthetic charm it possesses must be incidental. But a musical development also obeys laws; it is not capricious. And although these laws are doubtless more various than those of logic and have not yet been isolated and given names, they nevertheless exist.

And in neither case is the beauty of the development some kind of extraneous quality mysteriously added on. It is in virute of the fact that the development does obey laws that it is beautiful. Here, at any rate, we see some sense in Keats's remark that Truth is Beauty. When we come to an activity, such as physics, or a great part of literature, which is concerned with matters of fact, the notion of truth becomes somewhat different. Here we have the criterion of correspondence with an external world which need not have been as it is. But here, again, the truth is beautiful. Everywhere we encounter what Einstein calls "pre-established harmonies," and since the discovery and revelation of such harmonies is the concern, in their different regions, of both the artist and the man of science, we see once again that there is no essential distinction between the sciences and the arts.

In discussing the æsthetic aspect of science we have been discussing, perhaps at too great length, an aspect which appeals to the layman hardly at all. The layman is impressed, naturally enough, by the magnificent panorama of nature that science spreads before him, but the beauty of scientific theories is perhaps a consideration that appeals only to the scientific man.

The matter is different, however, when we come to what may be called the *moral* values incorporated in science. These have always been held, theoretically, in the greatest respect, especially the chief of them, namely, disinterested passion for the truth. But, as we have said, if we are to judge by its comparative rarity in the other activities of mankind, it is the most unpopular of virtues. How does it come about, then, that it plays so dominant a role in science? The first, and most obvious, reason is that science is an activity where success is not possible on any other terms. A business man, by concealment and misrepresentation, may become rich, that is, he may, by his standards, achieve success. A politician who was impatient with misleading catchwords, who really tried to think things out, would probably find his "usefulness" destroyed, since he might become incapable even of simulating that degree of conviction and moral fervour which is necessary to sway large audiences. And an advocate whose speeches should reflect the purely scientific attitude, giving every fact its true weight, would not be likely to have a very successful career. Even those advocates who, we are told, never embrace a cause of whose justice they are not convinced, are hardly scientific, since they evidently have an extraordinary capacity for arriving at definite conclusions on matters which are obviously debatable.

Science is the activity where truthfulness is most obviously an essential condition for success. Its success, in fact, is measured by its truthfulness. Of hardly any other human activity can this be said. In nearly everything else truth is a means and not an end. And if it turns out to be an unsatisfactory means, it is quite natural that it should be replaced by something else. But a scientific man who should misrepresent his observations, or deliberately concoct arguments in order to reach false conclusions, would merely be stultifying himself as a scientific man. He would not be prosecuting science. An advertisement may tell lies, but then telling the truth is not its object. Its object is to sell the stuff, which is an entirely different object. This is not to say, of course, that scientific men invariably tell the truth, or try to, even about their science. They have been known to lie, but they did not lie in order to serve science but, usually, religious or anti-religious prejudices. They were aiming at a different form of "success."

Such cases are, as it happens, very rare. Perhaps the reason is to be found in the fact that the success thereby achieved is very short-lived. Always the experiments are repeated, or the reasoning checked. The rigorous criticism, the complete lack of indulgence, that is shown by the scientific world, is one of its most agreeable characteristics. Its one simple but devastating criterion,

"Is it true?" is perhaps the chief characteristic that makes it seem such an oasis for the spirit in the modern world.

One reason, then, for the embodiment of the disinterested passion for truth in science is that the activity is meaningless without it. But how did men come to pursue an object which can only be reached by what seems to be generally considered as a very painful discipline? The answer is that, owing to the nature of the subject-matter of science, the discipline was not felt to be painful. On all matters where their passions are strongly engaged, men prize certitude and fear knowledge. From certitude can come purpose and a feeling of strength. It breeds courage and action, and is a ready means of ensuring that most desired of all things, an increased sense of vitality. Only the man of strong convictions can be a popular leader of mankind. For most men in most matters, whether it be the justice of a war, the rightness of a political creed, the guilt of a criminal, the wholesomeness of apples, certitude, in the entire absence of adequate evidence, is easily arrived at and passionately welcomed.

The scientific insistence on evidence, and the scientific absence of generosity in drawing conclusions from evidence, are resented in these matters. The scientific attitude, it is felt, with its promise of a long and very probably inconclusive investigation, would merely dam up the emotions that are clamouring for an outlet. But the matters with which science has hitherto been chiefly concerned are comparatively indifferent to us. For that reason science has been so successful. When Galileo investigated the law governing the motion of falling bodies, we cannot imagine that he cared in the least what the law would turn out to be. He could search for the truth with a single mind because none of his emotions could be outraged by the result. Similarly Newton's demonstration of the law of inverse squares roused no horror anywhere. Nobody had a strong emotional preference for the law of inverse cubes.

Towards most of the results of science we are indifferent. Their charm lies in the fact that they illustrate a harmony, but the results, in themselves, are matters of indifference. Any other results would do, provided they illustrated an equally beautiful harmony. The empirical fact that the velocity of light is nearer to 186,000 miles per second than it is to double that figure, excites no particular interest. But the fact that there must be an unsurpassable critical velocity in the sort of non-Euclidean universe we live in, is a matter of great interest.

Our reaction to most scientific facts is one of indifference, and our reaction to most scientific theories is one of purely æsthetic appreciation. But when a scientific theory has a philosophic, religious, or briefly, a "human" interest, we find at once that we are no longer content with our role of the disinterested seeker after truth. The opposition encountered by the Copernican theory and the Darwinian theory gives sufficient evidence of this. The splendid moral integrity manifested in scientific work, therefore, is due very largely to the nature of scientific material. It shows us the height to which man can rise, provided that a part only of his nature is involved. Science is truthful because it has practically no temptation to be anything else. In his work the scientific man is an artist, and his moral standard is superb, but the value of his example to the rest of mankind is limited by the fact that, in his work, the scientific man is not completely a man.

CHAPTER VIII

Towards the Future

THE ultimate justification of any intellectual activity is, it appears, its effect in increasing our awareness or degree of consciousness. Increase of consciousness appears, too, to have been one of the purposes of evolution, if we are to attribute purpose to that process. Certainly the most significant factor in the development from amœba to man seems to us to have been the increase in consciousness. Also the activities we most value are those that do the most to increase our awareness, of ourselves, of our fellow-creatures, and of the material universe we live in.

This is obviously the case with art. The great artist, painter, poet, or musician, makes us aware as we have never been aware before. He extends and subtilizes certain elements of our experience and so gives us greater knowledge and mastery of life. It is even possible that he acquaints us with radically new experiences and, if he be a great artist, we feel that these experiences are not freakish, but significant because in the main line of man's development. He voices:

> The prophetic soul of the wide world
> Dreaming on things to come.

This is the life-giving quality of art, and the added comprehension so bestowed is, as we have said, an essential element in what is called the æsthetic emotion.

The great value of science, from this point of view, is that it has introduced us to new ways of thinking. This is particularly true of the mathematical sciences. Huxley's statement, that scientific thinking is "organized common sense," is most plausible when applied to such a science as his own, biology, but it sounds very strained when we think of the new forms of the quantum theory and of the non-Euclidean geometries. Here we are in the presence of imaginative efforts which are altogether outside what is called common sense.

In the non-mathematical sciences new ways of thinking are not so apparent. Indeed, it may be through the limitations of common sense that these sciences are in their relatively unsatisfactory condition. Biology, for example, appears to be in urgent need of new concepts, and it does not seem that any biologist has yet appeared who is capable of the requisite imaginative effort. The present-day difficulties of physics itself probably spring from the fact that its imaginative efforts have not been imaginative enough. We are still hampered by our habitual modes of thought even when, as with the modern mathematical physicist, they have departed a long way from common sense. As Professor Lindemann says, when pleading for the necessity of a revision of our notions of space and time:

> Nevertheless, it is impossible for us, constituted as we are, to escape from spatio-temporal co-ordinates. We cannot think in other terms, we cannot even speak the new language which would be required. Physics expressed in quanta of action would convey nothing to our minds. We are therefore obliged to submit to the necessity of describing reality in terms of space and time, we are compelled to use unsuitable, and in certain circumstances meaningless, co-ordinates. All the quatum difficulties arise from the fact that we have not recognized the limitations of the spatio-temporal description of ultimate units. Once this has been fully apprehended, the results appear natural and inevitable.

We may say that the chief part played by science in helping on the developing consciousness of man is to be found in the new thoughts that it has made us think. The great merit of Einstein's theory, for example, from this point of view, is that it has enriched our consciousness by making us acquainted with a new and valuable way of thinking. This would remain a valid acquisition even if the theory, as a physical description of the universe, were disproved tomorrow. It would still be valuable for its effect in heightening our intellectual discrimination and subtlety, just as a work of art can be valued for its effect in refining and subtilizing our emotions.

But besides this cultural aspect of science, which is

perhaps only to be fully appreciated by the student, science has aided the growth of consciousness by making us more aware of the universe we live in. It has had, on a great scale, the cultural effect popularly attributed to foreign travel, and, like foreign travel, it can profoundly disturb some of the convictions with which we were quite content at home.

Amongst the ingredients of our total outlook that must be attributed to science is our modern conviction that natural phenomena behave in an orderly manner. That nature is orderly is a conviction that was held before modern science was born, but it was then based on beliefs about the nature of God and the destiny of man. The conviction is now based on entirely different grounds—chiefly on the work of Isaac Newton.

The belief that nature is orderly is not yet universal. Savages, we are told, live in a completely capricious universe, and we still find congregations praying for rain although they would hesitate, probably, to pray that the sun might stand still. That is because astronomy is a more developed science than meteorology. As we know, this belief that every natural process is determined by strict causality is not now held by all men of science. But the scepticism only affects the minutiæ of nature. For all the phenomena that fall within ordinary experience determinism holds. Astronomers continue, with complete confidence, to calculate eclipses.

Science, besides convincing us that nature is orderly, has expanded our imaginations by giving us utterly new and unsuspected information about the scale of the universe. It is difficult to say which is the more alien to our ordinary experience, the upper limits or the lower limits of the universe, that is, the distances and times dealt with by astronomy or the distances and times dealt with by physics. Perhaps the astronomical distances are the most alien for, as Eddington has pointed out, a man's physical dimensions are about half-way between those of an atom and those of a star, but the smallest of physical entities, an electron, is not so much smaller than an atom as a star is smaller than the whole universe. This last ratio, of course, only has a meaning if the universe is finite, and this is what the modern theory supposes it to be.

Thus, in spite of its measured distances of millions of light years, the modern universe is a smaller thing than the infinite space, with its infinite number of galaxies, dreamt about by such a man as Bruno. Yet there is a sense in which it is difficult to say even this. For this finite universe is supposed to be expanding, and expanding with an ever-increasing velocity. Thus no permanent limit can be assigned to it. It appears to remain finite, however, in the sense that only a finite amount of matter continues to populate it.

The total amount of matter in the universe can be tentatively estimated on theoretical grounds. The estimates vary, but a recent one given by Eddington is at least easy to remember. All the bodies of the universe, as we know, are collected into galaxies—the spiral nebulæ, or "island universes," as they are sometimes called. Eddington supposes that there are about one hundred thousand million of these galaxies scattered throughout space, and that each galaxy contains, on an average, one hundred thousand million stars. Observation halts a long way behind this estimate. Our telescopes reveal to us only about one part in a hundred thousand of this total.

Amongst the galaxies that have been observed, the most remote of those whose distances have been measured has a distance of one hundred and fifty million light years. This is receding from us with a velocity of fifteen thousand miles per second. Practically all the galaxies whose motions have been measured are found to be receding from us, and it is probable that the few which do not share in this motion will be found to do so when the proper corrections for the motion of the sun have been applied. This is, of course, what we should expect on the theory of an expanding universe. We can hardly suppose that this general recession is a mere coincidence, that the random motions of the galaxies are such that they all happen to be going away from us. But besides this qualitative confirmation, the measured velocities of recession confirm the theory quantitatively. For they recede faster the farther they are away, which is exactly what the theory predicts.

We have already remarked on the fact that this theory is a theory of *expanding space*. So far as the motions of the galaxies are concerned it is obvious that we could picture them as moving in an infinite, Euclidean space —the space of common sense. But the theory which predicts these motions cannot assume such a space; it assumes the spherical space of relativity theory, and the fact that it predicts the motions is therefore a strong support for the truth of its assumption. The observed fact of the recession of the galaxies is perfectly compatible, of course, with the notion of an infinite Euclidean space. But the only theoretical reason yet given for the observed fact involves the idea of an expanding spherical space.

The fact that space is now expanding inevitably raises the question as to how this state of affairs came about. There are various answers to this question, and they are all of them highly speculative. One theory, favoured by Eddington, pictures the universe as originally consisting of a uniform distribution of matter throughout spherical space. Such a universe could endure in that state for a very long time, for it has a large element of stability. But local irregularities would lead gradually to the formation of condensations, and then an element of instability would have crept into the world. Between the condensations there would be a force of repulsion. There would also, of course, be the force of gravitational attraction. It would be possible for these forces to be perfectly balanced. But the condition is essentially unstable. The slightest disturbance would start the universe either contracting or expanding. Whichever tendency prevailed, would go on prevailing. For as the condensations receded the gravitational force between them would progressively diminish and so become less and less able to overcome the repulsive force. And in the opposite case the gravitational force would become more and more triumphant. As a matter of observation we conclude that the repulsive force has triumphed, and we conclude from theory that it will go on doing so.

This repulsive force pertains to everything in the material universe. It exists between the bodies composing a galaxy, and between the atoms of those bodies. But in these cases its influence is masked by the gravitational attraction. This fact removes an ambiguity that might otherwise attach to the term "expanding space." For it might be thought that the term meant that every pair of points in space shared in the expansion. Thus we might suppose that a foot-rule, a human body, and every other material structure shared in the expansion. The theory would thus be deprived of all meaning, for such a uniform expansion is meaningless. But the theory does not assert this. A foot-rule remains a foot-rule. It is only the inter-galactic distances that increase, and they increase in the perfectly intelligible sense that it would take more foot-rules to fill them. Even the galaxies themselves do not increase in size. Distant as are the stars composing a galaxy from one another, the gravitational attraction is yet too strong.

The universe is expanding relative to the standard metre. It is possible, of course, to view the matter the other way round, and to suppose that the metre is shrinking. To quote Eddington: [1]

> Let us then take the whole universe as our standard of constancy, and adopt the view of a cosmic being whose body is composed of intergalactic spaces and swells as they swell. Or rather, we must now say it keeps the same size, for he will not admit that it is he who has changed. Watching us for a few million years, he sees us shrinking; atoms, animals, planets, even the galaxies, all shrink alike; only the intergalactic spaces remain the same. The earth spirals round the sun in an ever-decreasing orbit. It would be absurd to treat its changing revolution as a constant unit of time. The cosmic being will naturally relate his units of length and time so that the velocity of light remains constant. Our years will then decrease in geometrical progression in the cosmic scale of time. On that scale man's life is becoming briefer; his threescore years and ten are an ever-decreasing allowance. Owing to the property of geometrical progressions an infinite number of our years will add up to a finite cosmic time; so that what we should call the end of eternity is an ordinary finite date in the cosmic calendar. But on that date the universe has expanded to infinity in our reckoning, and we have shrunk to nothing in the reckoning of the cosmic being.
>
> We walk the stage of life, performers in a drama for the benefit of the cosmic spectator.
>
> As the scenes proceed he notices that the actors are growing smaller and the action quicker.

---

[1] *The Expanding Universe.*

> When the last act opens the curtain rises on midget actors rushing through their parts at frantic speed, smaller and smaller. Faster and faster. One last microscopic blur of intense agitation. And then nothing.

The cosmic being will then presumably cease to think about us. But we, in the meantime, will have lived to the "end of eternity."

The hypothesis that the present state of affairs came about from an initial uniform distribution of matter is only one of a number of alternative theories. And there are some authorities who hold that the apparent recession of the galaxies is illusory. These velocities of recession are measured, of course, by analysing the light we receive from these galaxies, and measuring the shift of the spectral lines towards the red end of the spectrum. But this "reddening of the light," according to some authorities, does not necessarily indicate a velocity of recession. It may be accounted for, they suppose, by a gradual loss of energy suffered by the light on its long journey through the intergalactic spaces. But there is not, at present, any theoretical reason for this presumed loss of energy.

If the galaxies are not really receding, then it is still possible for astronomers to hold to the enormous age they have attributed to the universe. This age, based on various lines of evidence, works out at millions of millions of years. And if this age be accepted, the hypothesis of the annihilation of matter is, as we have seen, almost inevitable. For no other process but the actual change of matter into radiant energy could keep up the necessary supply of energy for the time required. But if the theory of the expanding universe be adopted, this vast lapse of time can no longer be allowed. It can be shown that, by tracing the process backwards, the universe would already have been concentrated into a compact mass long before we had explored millions of millions of years. The greatest age compatible with the theory of an expanding universe is one of some thousands of millions of years. This is not long enough, according to some authorities, to allow for the evolution of stellar systems having reached its present stage. On the other hand, the reduced time-scale does not make the annihilation of matter necessary. Enough energy to last for this length of time would be provided by the building up of hydrogen into the more complicated chemical elements. We have no direct evidence either for the annihilation of matter or for the building up of matter, so that, on this ground alone, there is nothing much to choose between the theories.

Both the size and the age of the material universe are at present matters of great uncertainty. In neither respect is the universe infinite. Although the spaces and the times dealt with by astronomy are vast beyond all imagination, they are not unlimited. We find the same character of finitude when we turn to the smallest distances and times that science has yet revealed. It is a curious fact that nature is limited in both directions.

We may say that the notion of infinity, useful as it is to the mathematician and philosopher, is not required by the physical investigator. The world of matter is, in all respects, a finite affair.

The smallest distances and times dealt with in physics are just as unimaginable as are the greatest distances and times dealt with by astronomy. In the facts it reveals, as well as in its theories, science has long outrun the possibilities of the pictorial imagination. This happens long before we reach the minutest entities with which physics is concerned. It happens, for instance, when we are told that the molecules of air, at ordinary temperatures, are colliding with one another five thousand million times a second. The number of molecules in an ounce of water is about a million million million million. The estimated wave-length of the highly penetrating "cosmic" radiation is about one-tenth of a million millionth of a centimetre. Minute as an atom is, the nucleus of an atom is about one hundred thousand times smaller. Round this nucleus the outer electrons are revolving several thousand million million times a second.

Such examples could easily be multiplied. But they do nothing to impress our imaginations, for we can form no pictures to correspond to them. The world of physical science, both in its lower and its upper limits, is something altogether alien to us. But in both cases there are boundaries. We have already seen that space is not infinite, but finite, although expanding. Similarly, minute as are the entities dealt with by physics, there is reason to suppose that there is a lower limit beyond which the notions of space and time measurements become meaningless. This is another aspect of a principle we have already mentioned, the principle of indeterminacy.

Implicit in the notion of a measured distance is the notion of an operation which could be carried out without disturbing that distance. In stretching a measuring tape between the walls of a building, for instance, we do not suppose that the pressure of the measuring tape on either wall disturbs the position of the walls. Yet, unless the walls are of infinite mass, the pressure does disturb them, although the error so introduced is, of course, altogether inappreciable. Even if we make our measurements with a theodolite, and thus use a ray of light as our measuring rod, the walls are still disturbed, for light exerts pressure. The very operation of measurement does in fact, introduce an error into the measurement, although in all practical and most scientific measurements, this error is altogether negligible.

But when we come to consider phenomena in which single electrons are involved, the errors so introduced are of the same order of magnitude as the quantities to be measured. And these errors can in no way be avoided. They are not due to human imperfections; they are necessary consequences of the very structure of matter and energy. The disturbance inflicted by our measurements cannot, even in the ideal case, be reduced to nothing. It must have at least the dimensions of the quantum constant "h"—an exceedingly minute quantity, but not nothing.

The fact is, that our notions of precise spatial and temporal location do, to some extent, break down when we come to consider the ultimate particles of matter. These notions are abstractions from gross experience. It is not so surprising, therefore, that they turn out to be not quite suitable to a different range of experience.

The sense in which the precise position and velocity of an electron is a meaningless concept has been well illustrated by Professor Lindemann. He says:

> All physical objects directly observable appear to have some colour. . . . One could imagine . . . an object which was bleached white the moment it was exposed to light. There could be no sense in discussing its colour. For the moment light fell upon it, i.e. the moment one endeavoured to ascertain its colour, it would appear white, and it would be meaningless to ask what its colour was in the dark. If all objects were of this nature, the concept of colour would never have been formed, and the idea that every object must have some definite colour would never have arisen.
>
> A well-known instance of a similar anomaly is the impossibility of describing the smell of chemical substances which are changed by being brought into contact with air. Nitric oxide in the presence of oxygen forms the well-known brown gas, nitrogen-

> peroxide. It is obviously meaningless to discuss the smell of nitric oxide. It could only be ascertained by bringing it into contact with the mucous membrane of the nose. It would be impossible to do this without its being exposed to air, i.e. converted into nitrogen-peroxide. Its smell therefore can never be ascertained, and there is no sense in discussing it.

The precise position and velocity of an electron could not be observed because it is disturbed by the very act of observing it. And this is true because all natural processes, including any conceivable act of observation, have an atomic constitution. Only those objects which are not appreciably disturbed by a single atom of energy, could, even ideally, be satisfactorily measured. We may remark here that this fact does not imply that a particular characteristic, say position, has a theoretical lower limit. We could, theoretically, determine the position of an electron as precisely as we pleased. But if we made the error of position zero, the degree of error imported into the electron's other characteristic, its velocity, would become infinite. And as we require to know both characteristics precisely before we can predict, that is, before we can apply the law of causality, we see that the law of causality is useless to us. It is useless because the information on which it must be based is, by the nature of things, inaccessible to us.

We have seen that the original elements in modern scientific thought, the new way of thinking, have come from the mathematical sciences. It is in these sciences that common sense has been found most inadequate. The other sciences, as chemistry and biology, have done relatively little in the way of making us acquainted with radically new ideas. It is true that there are some biologists who find the current common-sense outlook inadequate, but a satisfactory set of new ideas has not yet been evolved. The present break-away from long-established habits of thought owes practically nothing to the non-mathematical sciences. Their cultural value is to be found in their facts rather than in their principles.

Chemistry, for example, whose practical value to our civilization can hardly be exaggerated, is valuable more for the control it gives us over nature, than for the insight it gives us into nature. The chemist's power of control results, of course, from the insight he has obtained, but it so happens that great depth of insight is not necessary in order to gain great power of control. With a simple form of the atomic theory as his theoretical equipment the chemist can conquer vast regions of practice. The atomic theory, in order to be sufficient for chemical purposes, does not need to be very searching. Doubtless there are chemical phenomena, here and there, which the immensely complicated atom of the physicist will one day be required to illuminate, but for the most part a much simpler atom is sufficient for the chemist's needs. And the atomic theory is the one fundamental contribution that chemistry has to make towards our understanding of the structure of the universe. And it is only in its simple form, in its main lines, that the atomic theory is really helpful to the chemist. For this reason we are not justified in expecting great theoretical advances from the chemistry of the future. The direction of any possible theoretical advance in chemistry lies, so far as we can see, through physics. The present and probable future value of the science of chemistry to man rests almost wholly upon its immense practical importance.

Biology, perhaps the least developed of the studies that truly deserve the name of science, is in a very different case. Its practical importance, although very considerable, is not at present overwhelming. A massacre of chemists would bring about the downfall of our civilization; a massacre of biologists would have unfortunate, but not disastrous, results. Biology is not to be valued chiefly for its practical importance. Neither is it of any great consequence theoretically. It is seldom that we come across a biological idea that impresses us as a staggering intellectual conquest. Even the theory of natural selection does not reveal hitherto unsuspected resources of the mind. Indeed, it is difficult to resist the impression that, on its theoretical side, biology is exceptionally inadequate. It seems to be quite surprisingly deficient in distinctive concepts.

Those who have tried to introduce some sort of comprehensible order into the bewildering and multitudinous phenomena of life by constructing concepts peculiar to living systems have not been successful. So that, as things are, biology's main contribution to our theoretical understanding of the world is the stale and unlikely surmise that a living organism will turn out to be nothing but a mechanical system. But perhaps we should regard this not so much as a conviction as a sort of mental marking time while waiting for a more satisfactory idea to occur to somebody.

The chief cultural value of biology is not to be found either in its practical or theoretical aspects, but in the amazing panorama it spreads before us. There is no study whose subject-matter is so varied. Even the range in mere size is enormous, "a big tree being as many times larger than a small bacterium as the sun is larger than the tree. It is startling to find that there exist adult insects, with wings and legs, compound eyes, and striated muscles, smaller than the human ovum; that jelly-fish may reach nearly a ton in weight; that the largest elephant has clearance top and bottom inside a whale; or that there are protozoa larger than the smallest vertebrates." [1]

And the varieties in structure and habit of living things is no less remarkable than their variations in size. It is due to this enormous variety that biology is so rich in special problems. In this respect it is vastly more complicated than any other science. The total number of substances known to the science of chemistry, for example, is as nothing compared to the number of species of living things. And each species presents

[1] *Animal Biology,* Haldane and Huxley.

many more aspects to be accounted for than does any chemical compound. Indeed, the scientific account of living things does already, in its first stages, completely exhaust the science of chemistry. That unit of living organisms, the cell, both in the substances that compose it and in the arrangement of those substances, presents problems which extend beyond the frontier of present chemical and physical knowledge. And after that come all the problems that are peculiar to biology.

The sheer variety of the subject-matter, as we may well suppose, would make the discovery of general principles, admitting of no exceptions, one of immense difficulty. We apparently cannot even generalize the statement "all men are mortal" into "all living things are mortal," for it appears that the amœba eludes this law. It is not surprising that there are very few general laws in biology, and that not one of these laws can be convincingly shown to have a universal application. The doctrine of natural selection gives a neat explanation of a large number of phenomena, but there are also plenty of phenomena where it only succeeds by being stretched and modified in a not very convincing fashion. Similarly, the Mendelian theory of inheritance still encounters anomalies. We may liken these biological theories, at best, to the elastic solid ether theory of light. That theory also explained a great number of phenomena, and even succeeded in predicting new ones which were subsequently observed. But there were always details that remained recalcitrant, and although some of them yielded in time to successive modifications of the theory, the theory was never enabled to cover the whole field. We know that the fate of this theory was to be scrapped, lock, stock, and barrel. But it is quite possible, of course, for a theory to encounter similar difficulties and still emerge triumphant. It may be that the current biological theories will prove to be of this kind. It is more exciting, however, and probably more reasonable, to suppose that biology will give birth to radically new conceptions.

Judged by the austere standards of the mathematical sciences, biology is the most immature of the sciences, for such studies as psychology, sociology, and so on, are hardly sciences at all. The great intellectual conquests of the near future, the next great step forward in the development of the human consciousness, may be expected to come from biology. Physics looks as if it is approaching finality in the sense that it is approaching a stage of complete unification. Maxwell's theory unified light, electromagnetism, and radiant heat. The theory of relativity has effected a further great unification, so that gravitation and the laws of mechanics are also included within the scheme. So far as what is called "field" physics is concerned, we seem to be within sight of the end. The other great characteristic of the physical universe, its atomic constitution, is still somewhat strange to us, but the quantum theory seems to represent it correctly in its main lines. It seems very unlikely that radically new phenomena, requiring radically new principles of interpretation, wait to be discovered.

All the main lines of physical investigation have been found to converge, and although the convergence is not yet complete, there is every reason to suppose that in the not very distant future it will be. Of recent years physical science has not discovered any phenomenon which lies right outside its scheme. When stars —the white dwarfs—were discovered having a density thousands of times that of platinum, the fact, although new in our experience, could readily be fitted into the theoretical scheme of physics. The theory of the atom showed itself quite competent to accommodate this new fact; it could, indeed, have foretold it. As an earlier example we may instance X-rays, which were fitted in quite readily to the electromagnetc theory of light. There are plenty of points where the physical universe still puzzles us, of course, but it seems unlikely that it will ever again completely bewilder us. It really seems as if the fundamental "constants" out of which the physical universe is composed are known: such constants as the electric charges and masses of the proton and electron, the quantum of action "h," and the velocity of light.

The connexions between these constants, what intimate relations they may have, is still, of course, a matter for research. Why, for example, should the proton have eighteen hundred times the mass of the electron?

But such questions, if answered, can only lead to a still greater unity within the present scheme of physics.

It is possible, nevertheless, that our outlook on the physical universe will again undergo a profound change. This change will come about through the development of biology. If biology finds it absolutely necessary, for the description of living things, to develop new concepts of its own, then the present outlook on "inorganic nature" will also be profoundly affected. For science will not lightly sacrifice the principle of continuity. The richer insight into the nature of living matter will throw the properties of dead matter into a new perspective. In fact, the distinction between the two, as far as may be, will be abolished.

This effort is already being made, of course, but from the other end. The present tendency is to reduce living organisms to mechanical systems. In view of what is now happening in physics itself, it does not seem likely that this effort will be successful. "Particle" physics, with simple location in space and time, has definitely proved itself inadequate. It is very probable, as Whitehead maintains, that the notion of particle will have to be replaced by the notion of organism. In order to avoid a break of continuity the notions of physics will have to be enriched, and this enrichment will come from biology. We can look forward to a further synthesis. The science of mind, at present in such a rudimentary state, will one day take control. In the service of the principle of continuity its concepts will be extended throughout the whole of nature. Only so will science reach the unity towards which it is aiming, and the differences between the sciences of mind, life, and matter, in their present form, will be seen to be unreal.

Index

Q162 .S85 1973 c.1
Sullivan, J. W. N. 100107 000
The limitations of science / b
3 9310 00045231 6
GOSHEN COLLEGE-GOOD LIBRARY